THE ABC OF ATOMS

BERTRAND RUSSELL

With an Introduction by
David Alexandre Ellwood

SPOKESMAN

First published in 1923
This edition, based on the Fourth Impression (1927),
published in 2024 by Spokesman Books
publishing imprint of the Bertrand Russell Peace Foundation Ltd
5 Churchill Park, Nottingham, NG4 2HF, England
www.spokesmanbooks.org
with support of the Atlantic Peace Foundation

Copyright © The Bertrand Russell Peace Foundation Ltd.
Introduction © David Ellwood
Cover portrait copyright © Hans Erni

All rights reserved. No part of this publication may be reproduced, stored
in a retrieval system or transmitted in any form or by any means electronic,
mechanical, photocopying, recording or otherwise, without the prior
permission of the publishers.

ISBN: 9780851249292

A CIP Catalogue record is available from the British Library

TABLE OF CONTENTS

CHAPTER		PAGE
	INTRODUCTION *David Alexandre Ellwood*	i
I	INTRODUCTORY	7
II	THE PERIODIC LAW	17
III	ELECTRONS AND NUCLEI	29
IV	THE HYDROGEN SPECTRUM	44
V	POSSIBLE STATES OF THE HYDROGEN ATOM	58
VI	THE THEORY OF QUANTA	69
VII	REFINEMENTS OF HYDROGEN SPECTRUM	81
VIII	RINGS OF ELECTRONS	95
IX	X-RAYS	106
X	RADIO-ACTIVITY	118
XI	THE STRUCTURE OF NUCLEI	133
XII	THE NEW PHYSICS AND THE WAVE-THEORY OF LIGHT	145
XIII	THE NEW PHYSICS AND RELATIVITY	155
APPENDIX: BOHR'S THEORY OF THE HYDROGEN SPECTRUM		172

Introduction

DAVID ALEXANDRE ELLWOOD

Castles in the Sky

Bertrand Russell remembered vividly where he was at that moment — it must have been one of those moments that struck many people that way — but not so deeply as it affected Russell. He was flying over the Alps from Rome to Paris when the pilot made the announcement:

Einstein est mort!

Russell's heart sank a 1000 feet. The date was 18 April, 1955, and for Russell the sense of loss was both personal and universal — a foreboding sense of doom for the entire world.

In many ways the story began with the book you are presently holding in your hands. A jewel of science writing in which Russell reveals the secrets of the atom with clarity and verve. As a public intellectual, Russell formed an in depth understanding of the latest scientific discoveries and saw the importance of communicating his insights to the public at large. In his age as well as ours, the people and their governments must not become estranged from the work of scientists lest fear and insecurity may guide them blindly toward disaster. Perhaps nobody understood this better than Russell, and

ABC OF ATOMS

he dedicated his immense skill to bridging every new chasm that might alienate people from power.

In *The ABC of Atoms* Russell tells us about the structure of atoms as they were understood in 1923. There had been two decades of startling revelations that ruptured the foundations of a scientific order that seemed both definite and eternal. The age of innocence was past, and the fundamental constituents of matter were revealing an inner complexity that few could fathom. The simplest atom, and the only atom that had proved amenable to theoretical analysis at the time, was Hydrogen, the lightest amongst all the elements. Still today, students are introduced to the theory of atoms by first working out the detailed structure of Hydrogen, and Russell does not pull any punches in explaining what we now call the Rutherford-Bohr model of the atom and the Bohr-Sommerfeld quantisation rules. This theory was remarkably successful in explaining the hydrogen spectrum as well as some properties of *hydrogen-like* systems. Russell's book provides a fine introduction to these subjects, infused with the additional thrill of reading science in the making. For the story which Russell tells of the atom was far from complete, but it's one that is indispensable reading for anyone who would like to trace the wild century of discovery that was to follow. Indeed, clarifying his ideas and writing this book certainly afforded Russell a special vantage point from which to view the scientific adventure that was unfolding before his eyes, but unseen to almost everyone else at that time.

After the terrifying use of nuclear weapons to bring a sudden end to the war with Japan, a grim foreboding hung over the new world peace concerning what such weapons would mean for the future of the world. And so it was that Bertrand Arthur William Russell (3rd Earl Russell) addressed the House of Lords of the United Kingdom at just after 4 o'clock on Wednesday afternoon, November 28, 1945. It was the first time Russell had spoken at length to the upper house, but it was not to address the terrible tragedy that had befallen the inhabitants of Hiroshima and Nagasaki just 3 months prior. Rather, Russell saw penetratingly into the future thanks to his

INTRODUCTION

keen interest in atomic physics and the ominous possibilities nature allowed for weapons of even greater destructive capacity:

> ...I should like to begin with just a few technical points which I think are familiar to everybody. The first is that the atomic bomb is, of course, in its infancy, and is quite certain very quickly to become both much more destructive and very much cheaper to produce. Both those points I think we may take as certain. (...) There is a further point which perhaps relates to the somewhat more distant future. As your Lordships know, there are in theory two ways of tapping nuclear energy. One is the way which has now been made practicable, by breaking up a heavy nucleus into nuclei of medium weight. The other is the way which has not yet been made practicable, but which, I think, will be in time, namely, the synthesizing of hydrogen atoms to make heavier atoms, helium atoms (...). In the course of that synthesis, if it can be effected, there will be a very much greater release of energy than there is in the disintegration of uranium atoms. At present this process has never been observed but it is held that it occurs in the sun and in the interior of other stars. It only occurs in nature at temperatures comparable to those you get in the inside of the sun. The present atomic bomb in exploding produces temperatures which are thought to be about those in the inside of the sun. It is therefore possible that some mechanism, analogous to the present atomic bomb, could be used to set off this much more violent explosion which would be obtained if one could synthesize heavier elements out of hydrogen.[31]

Quite incredibly, the journey that began with *The ABC of Atoms* had led Russell to conceive of the possibility of a Hydrogen bomb — or thermonuclear weapon — just months after the great secret of atomic energy had been revealed to the world in the obliteration of two Japanese cities. But Russell was more than a prophet, he was a statesman for all humanity. In this maiden speech to his peers he already set himself about the task of precluding the inconceivable horror that might follow if steps were not taken to outflank the suicidal folly of a society blindly setting alight a fuse to its own destruction:

> *We do not want to look at this thing simply from the point of view of the next few years; we want to look at it from the point of view of the future of mankind. The question is a simple one: Is it possible for a scientific society to continue to exist, or must such a society inevitably bring itself to destruction? It is a simple question but a very vital one. I do not think it is possible to exaggerate the gravity of the possibilities of evil that lie in the utilization of atomic energy.*[31]

Russell went on to explain that the institution of war must ultimately be abolished. This was not naive, but a logical conclusion given that scientific progress had already furnished weapons that can destroy cities, and will no doubt endow men with an ever greater capacity to realize their total annihilation. The first step, Russell explained, must be an honest appraisal of the current situation, leading to a sincere and open dialogue about the future for the sake of all humanity. In achieving this, Russell suggests the *scientists* might play a special role:

> *I think one could make some use of the scientists in this matter. They themselves are extremely uneasy, with a very bad conscience about what they have done. They know they had to do it but they do not like it. They would be very thankful if some task could be assigned to them which would somewhat mitigate the disaster that threatens mankind. I think they might be perhaps better able to persuade the Russians than those of us who are more in the game; they could, at any rate, confer with Russian scientists and perhaps get an entry that way towards genuine co-operation. We have, I think, some time ahead of us. The world at the moment is in a war-weary mood, and I do not think it is unduly optimistic to suppose there will not be a*

INTRODUCTION

great war within the next ten years. Therefore we have some time during which we can generate the necessary genuine mutual understanding.[31]

Time was of the essence and, unfortunately, Russell's speech failed to imbue those in authority with the necessary courage to put the genie back in the bottle. The United Nations could not agree on international control of atomic energy and the USSR successfully detonated an implosion-type plutonium bomb on August 27, 1949. The arms race had begun.

On October 3, 1952, the United Kingdom tested its first atomic bomb in the hull of *HMS Plym*, a *river class* frigate that had seen extensive service on escort missions during the Second World War. *Plym* was moored about 400 metres from the island of Trimouille in the Montebello Islands, Western Australia. The device was another implosion-type plutonium bomb and realized the expected yield of 25 kilotons of TNT. The blast left a saucer-shaped crater on the seabed 6 metres deep and 300 metres wide, but nothing remained of the frigate save a few chunks of metal that fell like rain from the sky and a "gluey black substance" that washed up on the beaches nearby.[5] However, Britain's day in the sun was short lived when Russell's nightmare arrived the following month in the form of a monstrous US experimental H-bomb called *Ivy Mike*. The yield was a foreboding 10.4 megatons of TNT, an astonishing increase just as Russell had predicted. The Soviets soon countered with the test of a more modest H-bomb on August 12, 1953.

The peoples of the world remained in an uneasy slumber while their leaders softly whispered to themselves the grim fairytale that *'the annihilating character of these agencies may bring an utterly unforeseeable security to mankind.'*[7] This childish fantasy soon became untenable when the United States detonated its first prototype of a deliverable thermonuclear weapon at Bikini Atoll on March 1, 1954. The yield was an *unexpected* 15 megatons of TNT, three times what the designers had predicted, destroying much of the instrumentation set up to monitor the blast. *Castle Bravo's* fireball towered up into the sky, and within one minute of detonation reached a height of 14 km and a diameter of almost 11 km. After 10 minutes the mushroom cloud was 40 km tall and 100 km wide, and in its wake left 18,000 square kilometres of the surrounding Pacific Ocean contaminated, affecting inhabitants of the small islands of Rongerik and Rongelap in the Marshall Islands.

Castle Bravo was not only more powerful than expected, it was also dirtier, and a shift in the wind cast a great deal more exposure to radiation than had been predicted. The Americans were initially dismissive, but the test soon became an international incident when the crew of a Japanese fishing boat, the *Daigo Fukuryū Maru* (Lucky Dragon No. 5), arrived back in port with her 23 strong crew showing signs of serious radiation exposure. While fishing outside the official exclusion zone, they had been showered with the remains of coral that had been heaved up and transformed into a white powdery ash. The blast had gouged a crater nearly 100 meters deep and over 1.5 km wide, whose contents rained down on the *Fukuryū Maru* and neighbouring islands with pernicious effect. Embedded in the coral were highly radioactive *fission fragments* from the uranium *tamper* that surrounded the device. All the crew suffered acute radiation syndrome soon after the incident, and the the boat's chief radioman, Kuboyama Aikichi, died on September 23, 1954 from complications of radiation sickness. The remaining 22 crew members remained in hospital for 14 months, regaining their freedom on May 20, 1955, but forever cursed by the incident.

Man's ability to destroy had now surpassed a thousand times the hideous proportions wielded by the first atomic bomb on Hiroshima, and the universal concern about the H-bomb was addressed by the BBC's flagship news programme *Panorama*. The broadcast included debates on the military, strategic, political and ethical dimensions of nuclear weapons, the latter pitting Russell against the Archbishop of York. Of most interest to Russell was an introduction to the relevant science by Joseph Rotblat, a Polish-born British physicist who had worked on the Manhattan Project but resigned on moral grounds. The BBC Director-General hosted a dinner after the

INTRODUCTION

recording where Russell had the opportunity to converse with Rotblat in depth and was thoroughly impressed. In 1949 Rotblat had turned his attention to the medical and biological uses of radiation and took up a position as Professor of Physics at St Bartholomew's Hospital ("Barts"), London. After obtaining data from a Japanese professor on the coral ash that showered down on the *Fukuryū Maru*, Rotblat realized the actual fallout was 40 times greater than expected, and deduced *Castle Bravo* was likely a three-stage fission-fusion-fissio design.

Rotblat became Russell's go-to expert on nuclear matters, and as his knowledge deepened ever further, Russell realized he must take on the arms race as his primary concern. After some difficulty, he convinced the BBC to broadcast *Man's Peril* — a sombre 'dirge for the human race' in which he said 'exactly how dreadful the prospect was unless measures were taken.'[30] The programme was aired two days before Christmas 1954, and helped allay some of Russell's personal anxiety, knowing that he had finally 'found words adequate to the subject.' The public response was encouraging, and Russell was overwhelmed with letters and requests for speeches and articles. Two groups stood out as noteworthy. On the political side, the most promising were the *World Parliamentarians* and the *Parliamentary World Government Association*, who invited Russell to speak at a joint meeting they were organizing in Rome in April, 1955. The other group were the scientists, particularly the French nuclear physicist Frédéric Joliot, who praised Russell's efforts and urged his support for an international conference of scientists to independently assess the full ramifications of the atomic age. Joliot's initiative chimed well with Russell's own idea of mobilizing the scientific community, but his Communist affiliations precluded Russell from overtly supporting Joliot's efforts. Instead Russell returned to his original thought of engaging eminent scientists to affect an independent dialogue that might engender the essential trust and cooperation between nations. He began reworking the text he had written for *Man's Peril*, and simultaneously wrote to Albert Einstein to ask if he would be willing to cooperate. On February 11, 1955,

ABC OF ATOMS

Russell wrote:

> *In common with every other thinking person, I am profoundly disquieted by the armaments race in nuclear weapons. You have on various occasions given expression to feelings and opinions with which I am in close agreement. I think that eminent scientists ought to do something dramatic to bring home to the public and governments the disasters that may occur. Do you think it would be possible to get, say, six "scientists" of the very highest repute, headed by yourself, to make a very solemn statement about the imperative necessity of avoiding war? Those chosen should be so diverse in their politics that any statement signed by all of them would be obviously free from pro-Communist or anti-Communist bias.*

Russell went on to describe what he saw as the essential points. In particular he stressed:

> *The thing to emphasize is that a future war may well mean the extinction of life on this planet. The Russian and American governments do not think so. They should have no excuses for continued ignorance on this point. And although the H-bomb at the moment occupies the centre of attention, it does not exhaust the destructive possibilities of science (...) this reinforces the general proposition that war and science can no longer coexist.*

Einstein replied within the week offering his full support for the initiative. Russell's final letter to Einstein was dated April 5, 1955. In that letter Russell included a draft of his statement for Einstein to endorse. And so it was that Russell left the conference of World Parliamentarians in Rome, eagerly awaiting a response from Einstein when he boarded his flight to Paris on April 18, 1955. His whole plan depended on Einstein's signature, since Niels Bohr, whom Einstein had urged to assist Russell, had refused to show any interest. It goes without saying that Einstein was the most famous scientist in the

INTRODUCTION

world, and commanded a degree of respect — one might even say awe — amongst even his most accomplished peers. And so when Russell heard the news that Einstein had died, he felt shattered, fearing that his plan would fall apart without Einstein's support. But on arrival at his hotel in Paris Russell miraculously found a letter from Einstein awaiting him. He had arranged for his mail to be forwarded from London, and was no doubt ecstatic to find Einstein's response, dated April 11, 1955, just four days before Einstein's passing:

> *Thank you for your letter of April 5. I am gladly willing to sign your excellent statement. I also agree with your choice of the prospective signers.*

Einstein's signature on Russell's statement was to be his last. He became fatally ill just two days later. The statement, which is now known as the Russell-Einstein Manifesto, was released at a press conference at Caxton Hall, London on July 9, 1955. Russell recruited Joseph Rotblat, who was also the youngest signatory, to organize and chair the meeting. The room was packed to capacity, and Russell read out the manifesto which ends with the resolution:

> *In view of the fact that in any future world war nuclear weapons will certainly be employed, and that such weapons threaten the continued existence of mankind, we urge the governments of the world to realize, and to acknowledge publicly, that their purpose cannot be furthered by a world war, and we urge them, consequently, to find peaceful means for the settlement of all matters of dispute between them.*

The intervening years, between then and now, can hardly be described as peaceful. While we have survived a Cold War, the *Doomsday Clock*[1] ticks ever closer to midnight. While our future may be less certain than ever, we must remember agency exists only in our present.

---

[1] In 2023 the Science and Security Board of the *Bulletin of the Atomic Scientists* set their Doomsday Clock to 90 seconds to midnight—the closest to global catastrophe it has ever been.

ABC OF ATOMS

Atomic Structure

In the following I have attempted to give a brief overview of some of the most important scientific discoveries surrounding Russell's *The ABC of Atoms*. I have written the overview so as to be self contained, so you may read it now before reading Russell's book. However, my suggestion is rather that you dive directly into Russell's *ABC* and enjoy his elegant prose gently leading you through the fascinating labyrinth of the atom. Russell's approach to science writing is quite unlike the protestations of today. He asserts almost nothing, and instead builds up a narrative that explains almost everything about the subject through simple facts and deductions. Although aimed at the widest possible audience, his book is a serious one, and so if you chose to return here after reading the book, I hope you will find something of interest that might spur you on to read some of the recommendations for further reading I have listed at the end.

Peering into the Heart of Matter

The 5 July 1687 is a date that is dear to many physicists. For a glorious span of 200 years following the publication of *Philosophiæ Naturalis Principia Mathematica* on this day, physics advanced by subjugating all natural phenomena to the prescripts introduced in Newton's magnum opus. The *classical* period of physics succeeded in conjuring up the macroscopic world from a flux of unobserved submicroscopic *atoms* – indivisible particles[2] – about which virtually nothing was known save their motion obeyed Newton's laws. Such analysis provided a comprehensive and detailed understanding of the material world that would reveal profound insights into the ethereal phenomena of electromagnetism and heat. However, the discoveries of X-rays (1895), radioactivity (1896) and the electron (1897) unlocked an unparalleled period of experimental creativity and theoretical revolution that would eventually depose Newton, and usher

---

[2]*atom* derives from the greek "ατομος" which simply means *uncuttable*.

INTRODUCTION

in a new age whose conceptual interpretation defies almost every common sense notion evoked in the classical era.

The explanation of any particular physical phenomenon is an act of imagination, but such explanations as have been recorded since mythological times are mere flights of fancy unless they are simultaneously both tightly constrained and explanatorily expansive. In other words, to be "tenable", a theory must first explain the phenomenon at hand without invoking a mechanism that inadvertently suggests a great many falsehoods. To be "good", a theory must go on to make quantitative predictions, ones that can be tested in a laboratory and either verified or refuted. No matter how logically compelling or aesthetically appealing a theory may be, any misstep in the laboratory renders it as simply wrong, and the theory must be rejected, in whole or in part. Finally, to be "great", a theory must transcend its origins to reveal a hidden order that brings together an assortment of otherwise unrelated truths under the same rubric. Newton's theory of Universal Gravitation is a classic example of a great theory.[3] Quantum Mechanics is the archetypal theory of a new republic, but one not founded on evolution, but revolution.

Towards the end of the nineteenth century the hypothetical particles — which collectively behaved so uniformly in Newton's clockwork universe — began to make themselves known individually and their character was anything but Newtonian. The first *elementary particle*[4] to be identified was the *electron* (e), whose charge to

---

[3]Newton showed that the same laws that accurately describe the flight of a cannon ball miraculously explain the orbital motion of celestial bodies. But Newton had not simply refashioned Kepler's precepts in new clothes, rather his theory cast its dominion over the entire universe with astounding precision. When small anomalies from Newton's laws were observed in the orbit of Uranus, the French astronomer Urbain Le Verrier showed them to be exactly what one would expect from the influence of an outer planetary neighbour. He calculated the orbit of the unseen world and sent the coordinates to Johann Gottfried Galle, a young astronomer who worked at the new observatory in Berlin. Galle looked for the unseen protagonist the same night he received Le Verrier's letter and found Neptune within 1° of its predicted position!

[4]a fundamental/indivisible particle in the original greek sense of ατομος, i.e.,

mass ratio was measured in 1897 by JJ Thomson via experiments involving *cathode rays*[5]. Thomson discerned that these beams are made up of tiny negatively charged particles whose charge/mass ratio is thousands of times larger than that of the lightest ion[6]. Of course, Thomson did not yet know if this large value was due to the immensity of an electron's charge or the minuteness of its mass, but he boldly leapt to the remarkable conclusion:

> ...*we have seen in the cathode rays matter in a new state (...) in which the subdivision of matter is carried very much further than in the ordinary gaseous state: a state in which all matter — that is, matter derived from different sources such as hydrogen, oxygen, etc. — is of one and the same kind; this matter being the substance from which the chemical elements are built up.*[38]

After the discovery of the electron, the naive picture of atoms as hard, indivisible, electrically neutral spheres with certain physical and chemical properties was no longer tenable. The word "atom" came to refer to an irreducible unit of *chemical identity*, but what exactly are these "units" and how do they manifest their particular characteristics? In the first decade of the nineteenth century John Dalton noticed that chemical elements[7] always combine in definite proportions (by weight) when forming a given compound.[8] Dalton reasoned that if a chemical compound consists of hypothetical particles — now called *molecules* — these particles must themselves be made up from specific combinations of atoms whose masses could be discerned from the relative weights of the reactants. Dalton's

---

one with no discernible structure.

[5]Walter Kaufmann performed similar experiments with cathode rays at about the same time. Kaufmann's results were more accurate than Thomson's, but he did not claim to have discovered a fundamental particle.

[6]i.e., electrically charged hydrogen (H^+).

[7]pure substances that cannot be decomposed further by chemical means.

[8]e.g., oxygen and hydrogen always combine in a fixed ratio of 8:1 to form water. There can be no variation.

INTRODUCTION

original guesses were all incorrect, but through the later work of Joseph Louis Gay-Lussac[9] and Amadeo Avogadro[10] it became possible to write down dependable formulas for various molecules and an ordering of chemical elements based on their *atomic weight* was revealed, that is, the mass of a given species of atom relative to that of hydrogen. Of course, that actual weight of atoms was only measured much later, but the nature of the ordering was itself suggestive. With a few notable exceptions (e.g., chlorine), atomic weights were found to be approximately whole numbers and, in 1815, William Prout suggested that all chemical elements consist of multiples of one fundamental particle, most likely the hydrogen atom. We now know that Prout's guess was only partially correct. The hydrogen ion (H^+) — known as a *proton* (p) in this context — does indeed play a fundamental role, but it was the discovery of the electron that would provide the master key to unlocking the inner structure of atoms.

First note that if the electron is a common constituent of every chemical element, but accounts for only a tiny fraction of its mass, one might imagine an electrically neutral atom consists of matching numbers of *positively* charged protons (p) and *negatively* charged electrons (e). Even if this were correct, how might these constituents be arranged? The breakthrough came from JJ Thomson's protégé, Ernest Rutherford, who began investigating the nature of radioactivity while still working under his supervision. Rutherford analyzed the mysterious *rays* discovered by Henri Bequerel and Marie Sklodowska-Curie and classified them into three types, *alpha*, *beta* and *gamma*. He went on to show that the first two types, *alpha* and *beta*, are composed of particles emitted with energies millions of times greater than can be produced in ordinary chemical reactions. The *beta* particles (β^-) turned out to be nothing other than high

---

[9]Gay-Lussac found definite rules of combination (by volume) in the reaction of gases, e.g., water is formed by combining two volumes of hydrogen with one volume of oxygen.

[10]Avogadro had the idea that equal volumes of any gas at a given temperature and pressure always contain equal numbers of particles.

energy electrons, but the *alpha* particles (α) were more difficult to identify. In 1906, Rutherford found the charge/mass ratio of *alpha* particles was about half that of hydrogen ions, and he showed that α's are in fact doubly charged ions of Helium (He^{++}), an element which has an atomic weight of four. He put this discovery to work in Manchester, together with Hans Wilhelm Geiger and Ernest Marsden. They began probing the atomic structure of matter by targeting a narrow beam of *alpha* particles at pieces of thin metal foil. Rutherford vividly recalled the breakthrough as follows (quoted from [40]):

> *One day Geiger came to me and said, "Don't you think that young Marsden, whom I am training in radioactive methods, ought to begin a small research?" Now I had thought that too, so I said, "Why not let him see if any alpha particles can be scattered through a large angle?" I may tell you in confidence that I did not believe that there would be, since we knew that the alpha particle was a very fast massive particle, with a great deal of energy ... Then I remember two or three days later Geiger coming to me in great excitement and saying, "We have been able to get some of the alpha particles coming backwards..." It was quite the most incredible event that has ever happened to me in my life. It was almost as incredible as if you fired a 15-inch shell at a piece of tissue paper and it came back and hit you!*

In 1911, Rutherford concluded[32] that the world of atoms was nothing like that which had hitherto been proposed, but instead resembled a *micro-cosmos* of fascinating complexity, one that would turn out to be governed by some surprising new laws of change and chance. Rutherford imagined that each atom consists of a tiny version of our own planetary system with almost all its mass concentrated at the centre in an extraordinarily dense positively charged *nucleus*, about which a family of negatively charged electrons orbit — in the same way the planets in our own solar system orbit the sun. Rutherford calculated that the nucleus of a gold atom must

INTRODUCTION

be at least 10000 times smaller than the atom itself, and a statistical analysis of the proportion of α's scattered through various angles provided firm evidence for Rutherford's model.

Russell's book is an exposition of the structure of atoms based on Rutherford's model, but predates some vital discoveries that we can only begin to sketch. At the time of writing, physicists were grappling with the problem of the nature of the atomic nucleus. Following Dalton and Prout, we might suppose the nucleus of an atom consists of a dense amalgamation of hydrogen ions, i.e., *protons* (p). However, Rutherford's own results about the charge/mass ratio of *alpha* particles ($^4\text{He}^{++}$) was already at odds with such a theory. To achieve the correct charge/mass ratio, the idea that became popular was that the atomic nucleus must be some sort of conglomeration of both protons and electrons. For example, the structure of a Helium atom was thought to consist of two electrons in orbit around a nucleus of four protons amalgamated with two electrons. Evidence for this idea seemed to come from Henry G J Moseley's determination of the nuclear charge of several medium weight elements in 1913. Moseley found that the nuclear charges of every element he studied was nearly an exact multiple of the magnitude of that of the electron. Moreover, when ordered by atomic weight, Mosely observed that the nuclear charge of neighbouring elements differ by exactly one unit. For this reason, the nuclear charge of an atom is now dubbed its *atomic number* (Z), and gives both the place number of a given element in the *periodic table* as well as the number of electrons in orbit around its nucleus.

The fact that radioactive substances emit both *alpha* ($^4_2\text{He}^{++}$) and *beta* (e) particles, but never individual protons (H^+), suggested that α's might themselves be fundamental building blocks of the atomic nucleus. Russell correctly rejects this idea, preferring to avoid adding any new ingredients to the atomic recipe until experimental evidence dictates otherwise. But the *mistaken* belief that an atomic nucleus consists of a *mélange* of protons and electrons was popular when Russell wrote his book and would remain so until James Chadwick discovered the *neutron* (n) in 1932. We will come back to the structure of the nucleus later, but first it's important to say a little about the orbital structure of the atom and the ensuing development of quantum mechanics.

Spectroscopy

That a prism can produce a rainbow of colours was known since antiquity[11], but are these colours an artefact of the prism or somehow inherent within the sun's rays? In a series of experiments beginning in 1666 Isaac Newton showed that a prism does not colour sunlight, but rather *splits* the sun's rays into a *spectrum* of colours which can also be recombined into "pure white" light. As optical technology improved it soon became apparent that the bands of colour in the sun's spectrum are not spread out continuously, but contain both bright and dark lines that convey information about the chemical composition of the sun's atmosphere.

Any incandescent substance glows with a hue that is indicative of its composition, but the information contained within the mesh of colours becomes most apparent when the radiating atoms are not bound but essentially free of one another. In the laboratory this is achieved by studying spectra produced by gas discharge tubes at very low pressure.[12] Such spectra display sharply defined lines of colour, which when quantified show surprising regularities. Soon simple formulae were found relating the lines in the emission spectra for Hydrogen (Balmer – 1885) as well as many other elements (Rydberg – 1900; Ritz – 1908). Russell beautifully explains how these formulae relate the reciprocal of the wavelength ($1/\lambda$) of different spectral lines and permit the description of a seemingly complex assortment of colours in terms of a simple set of basic *terms*. What these formulae meant did not become clear until the appearance of Niels Bohr's seminal papers in 1913.[2]

---

[11] as recorded by the Roman philosopher Seneca in his *Naturales Quaestiones*.

[12] a fraction of the atoms of the gas are broken down into electron-ion pairs, forming a conductive plasma that radiates at definite frequencies.

INTRODUCTION

The fact that the reciprocal of the wavelength, rather than the wavelength itself, simplifies the formulae for spectral lines hints at the fact that radiant energy is itself *quantised* in discrete chunks called *photons*. Back in 1905, the same year Einstein introduced his famous equation $E = mc^2$, he also introduced the less well known but equally profound $E = h\nu$. This equation is every bit as revolutionary as its more famous cousin relating energy and mass. Here "E" denotes the energy of a *photon* (considered as a "particle" of light) with its frequency ("ν") considered as an electromagnetic wave, where the constant of proportionality is *Planck's constant* (h). How a photon can simultaneously embody both *wave* and *particle* aspects was one of the early riddles of *quantum theory*, but the apparent paradox is merely an artefact of taking too naive a picture of the electromagnetic field. We shall not explain this here, but you should beware of much popular writing on this topic — the *quantised electromagnetic field* provides a model for radiation theory which fully synthesizes both its wave and particle aspects.

After learning about Rutherford's model of the atom, Bohr focused on the problem of explaining the relationship between spectral lines in terms of the orbital dynamics of electrons. Rather than trying to explain the observed frequencies from the periods of oscillating electrons, Bohr flipped conventional wisdom and instead regarded atomic electrons as restricted to certain "allowed" orbits called *stationary states*. A moving charge should radiate away its kinetic energy when bound in orbital motion, but in these special states Bohr postulated that the conventional laws of electromagnetic theory no longer apply and the orbital energy of an electron remains constant. According to Bohr, a given stationary state can only be altered when an electron gains or loses the specific quantity of energy necessary to *jump* between a pair of allowed orbits. In such transitions radiation is absorbed or emitted in the form of a photon whose frequency is given by the difference between the initial and final states of the atom according to Einstein's formula:

ABC OF ATOMS

$$\nu = \frac{E_{\text{photon}}}{h} = \frac{1}{h}(E_i - E_f)$$

The discrete nature of emission spectra, i.e., that radiation from a given species of atom is restricted to a set of definite frequencies, is thereby related to the existence of a discrete set of (circular) orbital states labelled by an integer parameter n (called a *quantum number*). Bohr was able to calculate[13] the energies of these *stationary states* for a one electron atom and show how they correspond with the empirically derived "terms" used in the Rydberg-Ritz formula. Moreover, Bohr's model also predicted the energies of the innermost orbits of heavier atoms, and was instrumental in Moseley's analysis of X-ray spectra and his determination of the nuclear charge mentioned above. This was a remarkable achievement and Russell explains Bohr's theory in detail, as well as its extensions and adaptations, most notably by Sommerfeld where electrons move in a more general set of *elliptical* orbits.

Quantum Theory

Despite all its success, the Bohr-Sommerfeld theory was limited in applications to one-electron-*like* systems and based on *ad hoc* assumptions that contradicted key ideas in classical physics. Why don't electrons bound in stationary states radiate away their energy and collapse into the nucleus? There was no way to understand why classical physics should apply in some circumstances but not in others, let alone derive the "quantisation" rules on which the theory depended. Russell's book was written on the cusp of a fascinating moment in history that would see the very foundations of classical physics swept away in a tremendous new *wave* of understanding.

---

[13] Bohr guessed that the *angular momentum* of an orbiting electron must be restricted to integer multiples of some constant $\hbar$. He then showed that $\hbar$ must be equal to Plank's constant h divided by 2π via the *correspondence principle* — i.e., that quantum and classical physics should correspond in the limit of *large n*.

INTRODUCTION

If the discovery of quantum mechanics was a play in four acts, Russell's book concludes at the end of the first act[14], i.e.,

Act I Niels Bohr shows that when considering the atom it is necessary to view light — which was *known* to be an *electromagnetic wave* — to consist of elementary particles called *photons*.

I emphasize *known* here since the debate on the nature of light was settled more than a hundred years earlier by the British polymath Thomas Young in a famous experiment. In 1801 Young showed that light exhibits constructive and destructive interference — a property of waves, not particles — in which the crests and troughs of coincident waves either amplify or cancel one another. Then, in 1864, Maxwell was able to identify light as a travelling disturbance in electric and magnetic fields, i.e., an *electromagnetic wave*. Bohr's theory confounded physicists because it flew in the face of everything we knew about light and electromagnetism, but it seemed Newton's theory of motion was left partially intact, at least in the description of certain "allowed" orbital states. The next phase of the quantum revolution was to turn the fundamental concept of motion on its head, indeed, the very conception of a *particle* or *atom* had to be abandoned before new dynamical laws applicable to the scale of atomic and subatomic physics could be discovered.

Act II Louis de Broglie shows that when considering the atom it is necessary to consider particles in motion as waves.

In his memoire the French physicist and aristocrat Louis de Broglie wrote:

> ...*After the end of World War I, I gave a great deal of thought to the theory of quanta and to the wave-particle dualism* ... *It*

---

[14]I have omitted many discoveries and taken liberties with history in order to introduce the reader of Russell's book to a few ideas in quantum mechanics. The interested reader should go on to consult the references given at the end of this introduction, in particular Gamow's wonderful little book [17].

> *was then that I had a sudden inspiration. Einstein's wave-particle dualism was an absolutely general phenomenon extending to all physical nature ...The new concept also gave the first wave interpretation of the conditions of quantizing the momenta of atomic electrons.*[26]

While still a student, de Broglie realised[10] that Bohr had only dealt with half of the problem when formulating his theory of the atom. If understanding atomic structure required waves (light) to be treated as particles (photons), then perhaps it is also necessary to treat particles (electrons) as waves? This idea suggested an immediate remedy to one of the central mysteries in Bohr's model. If the electrons orbiting a nucleus could be considered as waves, then their orbits would be restricted to certain *stationary states* in the same way that the notes produced on a guitar string are restricted by its length, i.e., only certain *standing waves* can fit into a given length of string because any such wave must consist of an exact number of half wavelengths. In the case of an atom, de Broglie concluded that the radii of allowed orbits must be restricted to values such that their circumference be an exact multiple of an electron's wavelength (λ_e):

$$2\pi r = n\lambda_e$$

But what is the wavelength of an electron? According to the description of light as photons, the magnitude of the momentum (p) of an individual photon is inversely proportional to the wavelength of light (λ) it composes. The constant of proportionality is again *Planck's constant h*. If this relation also applies to atomic electrons, then

$$p = \frac{h}{\lambda_e} = \frac{nh}{2\pi r} = \frac{n\hbar}{r}$$

On substituting the usual formula for momentum ($p = mv$) we re-

INTRODUCTION

cover Bohr's condition[15] for the quantisation of angular momentum

$$mvr = n\hbar$$

What is more, the scale of de Broglie wavelengths provided the first indication of when a classical description of motion will fail, and a new quantum dynamical law is needed. In 1925, the year after the publication of de Broglie's thesis, Clinton Davisson and Lester Germer accidentally discovered a diffraction pattern in the scattering of electrons by crystal surfaces[9], spectacularly confirming de Broglie's ideas. In 1929 Louis de Broglie was award the Nobel Prize in physics for his *"discovery of the wave nature of electrons."*

Act III Erwin Schrödinger formulates his theory of *wave mechanics* and shows how it applies to the atom as well as more general systems, thus establishing the foundations of modern (nonrelativistic) quantum dynamics.

From 1921 to 1927 Erwin Schrödinger was a professor at the University of Zürich. Schrödinger and Peter Debye at the Swiss Federal Institute of Technology (ETH) organized a joint seminar for physicists at these neighbouring institutions. Felix Bloch recalls attending the seminar while still a young student[1]:

> *Once, at the end of a colloquium, I heard Debye saying something like: "Schrödinger, you are not working right now on very important problems anyway. Why don't you tell us some time about that thesis of de Broglie, which seems to have attracted some attention?"*
>
> *So, in the next colloquia, Schrödinger gave a beautifully clear account of how de Broglie associated a wave with a particle and how he could obtain the quantisation rules of Niels Bohr and*

---

[15]Sommerfeld's quantisation condition has a similar explanation, i.e., that in a stationary state the phase of the wave representing the electron must change by an integer multiple of 2π on completing each orbit.

ABC OF ATOMS

Sommerfeld by demanding that an integer number of waves be fitted along a stationary orbit. When he had finished, Debye casually remarked that he thought this way of talking was rather childish. As a student of Sommerfeld he had learned that, to deal properly with waves, it was necessary to have a wave equation.

Just a few weeks later he gave another colloquium which he started by saying: "My colleague Debye suggested that one should have a wave equation; well, I have found one!"

And then he told us essentially what he was about to publish under the title "Quantization as an eigenvalue problem" as the first paper in a series in Annalen der Physik. I was still too green to really appreciate the significance of this talk, but from the general reaction of the audience I realised that something rather important had happened, and I need not tell you what the name of Schrödinger has meant from then on.

Schrödinger sought an equation for which de Broglie's waves are the solution, and he published his findings in a series of papers[33, 34, 35, 36]. To introduce Schrödinger's equation I'm going to employ some mathematical terminology. This is unavoidable, but what Schrödinger achieved is considered one of the most important breakthroughs in the 20th century and so it would be amiss not to mention a few details. If you find the following incomprehensible, do not despair. I've included a list of excellent references at the end of this introduction that will help you get to grips with quantum mechanics as well as the many other important developments that followed in the wake of Russell's book.

To represent a wave mathematically physicists employ a function $\Psi(t, \mathbf{x})$ that describes the wave's form and propagation. At any moment in time (t) and point in space $(\mathbf{x})$, the value of the function $\Psi(t, \mathbf{x})$ is a numerical measure of the physical quantity that is doing the waving — be it the vibration of a guitar string, ripples on the surface of a pond, or variations in the strength of an electric field.

INTRODUCTION

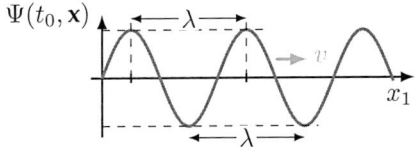
(a) Space function shows *wavelenght* λ

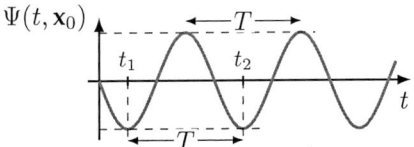
(b) Time function shows *period* T

Figure 1: A sinusoidal wave $\Psi(t, \mathbf{x})$

Figure 1 shows an example of a travelling wave that varies *periodically* in time and space as it propagates to the right[16] with speed $v = \lambda/T$. Here $\Psi(t_0, \mathbf{x})$ is the *space function* that captures a snapshot of the wave at a fixed instant of time, (t_0), while $\Psi(t, \mathbf{x}_0)$ is the *time function* that shows the waving at a given point in space $(\mathbf{x}_0)$.

If we focus on the special case of a standing wave, indeed any state of constant energy, the wavefunction can be split into factors so that the time and space dependency can be analysed separately. Thus for stationary states of the atom it is sufficient to consider the space function (at time $t = 0$) which we shall denote simply by $\psi(\mathbf{x})$. In the first paper[33] of the series Schrödinger introduced a time-independent equation for modelling the stationary states of the (non-relativistic) hydrogen atom and recovered Bohr's formula. For an electron of energy E in a Coulomb potential $V(\mathbf{x})$[17] Schrödinge

---

[16]here measured by the x_1-coordinate.
[17]$V(\mathbf{x}) = -Ze^2/r$, where Z is the atomic number ($Z = 1$ for hydrogen) and $r = \sqrt{x_1^2 + x_2^2 + x_3^2}$ in cartesian coordinates $\mathbf{x} = (x_1, x_2, x_3)$.

wave equation reads:

$$\left(-\frac{\hbar^2}{2m_e}\nabla^2 + V(\mathbf{x})\right)\psi(\mathbf{x}) = E\psi(\mathbf{x}) \tag{1}$$

Here ∇^2 is a differential operator called the *Laplacian*.[18] If you are not familiar with the mathematics involved, just try to think of this expression as an equation whose solution is the *wavefunction* $\psi(\mathbf{x})$. If we denote the expression in brackets by $\hat{H}$ then we can write this equation more succinctly as:

$$\hat{H}\psi(\mathbf{x}) = E\psi(\mathbf{x})$$

Physicists and mathematicians refer to this as an *eigenvalue problem*. A solution $\psi(\mathbf{x})$ is called an *eigenfunction* of the differential operator $\hat{H}$ with *eigenvalue* E. Thus the energy of the electron appears as the factor by which a wavefunction is rescaled — but otherwise unchanged — by the action of the differential operator $\hat{H}$.

If the electron is *bound* in an orbit, Schrödinger's equation only has solutions for certain values of E. Now a bound electron is one that cannot escape in any direction, and the associated *boundary condition* is that one must restrict attention to wavefunctions that vanish as each coordinate tends to infinity. In this way, without imposing any *ad hoc* rules, Schrödinger found a wave equation that has exactly n^2 solutions with energies

$$E_n = -\frac{Z^2 e^4 m_e}{2n^2 \hbar^3} \quad (n = 1, 2, \ldots)$$

and no other others, for any other energy! As Schrödinger himself remarked[33]

> *The essential thing seems to me to be that the postulation of 'whole numbers' no longer enters into the quantum rules mysteriously, but we have traced the matter a step farther back, and*

---

[18]In cartesian coordinates the Laplacian is given by the sum of second partial derivatives $\nabla^2 = \partial^2/\partial x_1^2 + \partial^2/\partial x_2^2 + \partial^2/\partial x_3^2$.

INTRODUCTION

found the 'integralness' to have its origin in the finiteness and single-valuedness of a certain space function.

Integralness, as Schrödinger puts it, arises in the same way as it does in the case of the node numbers of a vibrating string, and is no longer imposed by arbitrary rules as in the Bohr-Sommerfeld theory. Although Schrödinger's results are no different from those of Bohr, Schrödinger's theory has the advantage of incorporating the (experimentally verified) wave nature of matter into a general framework that also permits the analysis of more complex systems — systems for which Bohr's theory had proved wholly inadequate, e.g., multielectron atoms etc.[19]

In the fourth paper of his series[36], Schrödinger introduced a dynamical version of his wave equation to describe non-stationary states. In compact form the equation reads:

$$i\hbar \frac{\partial \Psi(t, \mathbf{x})}{\partial t} = \hat{H} \Psi(t, \mathbf{x}) \qquad (2)$$

Notice the wavefunction is now a function of both time and space ($\Psi(t, \mathbf{x})$), i.e., this equation can be used to predict how the wave properties of matter evolve with time. In particular, once we have found the stationary states of an atom by solving the eigenvalue problem described above, we may hope to use the dynamical equation to study how an atom transforms from one stationary state to another, e.g., when it emits or absorbs a photon. This problem goes beyond quantum mechanics because it involves the *interaction* of a charged particle (the electron) and the electromagnetic *field*, a full treatment of which requires the development of quantum theory in its most sophisticated form — *quantum field theory*.

Rather than attempt to say something about these more advanced aspects of quantum theory, let us instead draw attention to the appearance of the imaginary number $i = \sqrt{-1}$ in equation 2. Schrödinger had found that to incorporate time evolution into his theory he was

---

[19] indeed any dynamical system that admits a *Hamiltonian* formulation.

obliged to consider *complex numbers* rather than ordinary real numbers, and dynamical solutions $\Psi(t,\mathbf{x})$ are necessarily complex valued. Now we can look for stationary solutions to equation 2 as a special case, and we find any such solution takes the form:

$$\Psi(t,\mathbf{x}) = \psi(\mathbf{x})e^{-iEt/\hbar}$$

where $\psi(\mathbf{x})$ is an eigenfunction satisfying the time independent equation 1 with eigenvalue E. In this sense Schrödinger's dynamical equation 2 contains the time independent theory as a special case. By factoring stationary solutions in this way the space function $\psi(\mathbf{x})$ can always be taken to be real. Nevertheless, it was not at all clear to Schrödinger how to interpret solutions of his wave equation. In particular, in the case of an electron what is doing the waving? This question is even more serious in the case of dynamical solutions $\Psi(t,\mathbf{x})$. What can a complex valued wavefunction possibly mean?

Act IV Max Born uses Schrödinger's theory to analyse scattering experiments and provides the fundamental interpretive rule of quantum theory.

From the earliest days of Schrödinger's wave mechanics he and others puzzled over how to interpret the wavefunction $\Psi(t,\mathbf{x})$. After a conference dinner in 1926 Max Born recounts composing a short verse with his classmates[1]:

> *Erwin with his psi can do*
> *Calculations quite a few.*
> *But one thing has not been seen:*
> *Just what does psi really mean?*

The cold hard truth was that Schrödinger didn't know! At first he thought the wavefunction must represent the *shape of a particle*. Experiments had demonstrated that electrons exhibit diffraction and interference effects making their wave-like nature undeniable. Perhaps we have been wrong to use words like "particle" to describe

INTRODUCTION

atoms and electrons all along? The evidence seemed contradictory until Max Born employed Schrödinger's theory to model scattering experiments in which a particle (like an electron or alpha particle) strikes a target (like an atom) and ricochets off in a particular direction. If we envisage a particle as a *wavepacket* in the way Schrödinger had initially thought, i.e., as a wave shaped entity that is narrowly concentrated within a small region of space, Born showed that after impact the wavepacket loses its concentrated form and begins radiating out in all directions. Now an electron striking a target may scatter in any possible direction, but it never disintegrates and spreads out in all directions simultaneously.

Born resolved this apparent dichotomy with a completely new idea, that is to say, he suggested the wavefunction describes the *probability amplitude* of a particle rather than its smeared out shape in space[3, 4]. More precisely, Born proposed that the square of the magnitude of the wavefunction $|\Psi(t,\mathbf{x})|^2$ gives the probability that a particle is at (or very near to) the point "$\mathbf{x}$" at time "t".[20] In order to make sense of this mathematically, Born required that the wavefunction satisfy the *normalisation* condition[21]

$$\int |\Psi(t,\mathbf{x})|^2 d\mathbf{x} = 1$$

This is a major break with classical theory, even with the old quantum theory of Bohr and Sommerfeld described in Russell's book, and it is worth pausing for a moment to reflect just how far we have come:

- The dynamical picture of an electron orbiting the nucleus has been completely lost, and all we have in its place is a wavefunction $\psi(\mathbf{x})$ that determines a *static* probability distribution $|\psi(\mathbf{x})|^2$.

---

[20] i.e., located within a small volume element '$d\mathbf{x}$' centered at '$\mathbf{x}$'.
[21] which just means that we can be sure of finding the particle somewhere.

ABC OF ATOMS

This probabilistic view is completely consistent with experimental results, but is a major departure from the *exact* view of reality that classical theory had aspired to provide. Moreover, Born's postulate may have brought the wave and particle view points together in the laboratory, but it was far from congenial to the taste of many physicists. Whereas particles are always detected as localized entities in any one experiment, their behaviour in repeated experiments — or their interaction *en masse* — reveals a puzzling statistical pattern that can only be explained using wave mechanics and probability amplitudes.

Although Born's probabilistic interpretation of the wavefunction is empirically sound, it was hard for many leading figures to accept and Schrödinger, Max Planck, Albert Einstein and Louis de Broglie remained sceptical about it to the end of their days. There is much more we could discuss about this but let me finish here by mentioning that just before Schrödinger invented his theory of wave mechanics, a young German physicist named Werner Heisenberg had discovered a different approach to quantum phenomena that is now referred to as *matrix mechanics* [21]. Like Schrödinger's wave mechanics, Heisenberg's theory also reproduced Bohr's quantisation of the energy levels of the hydrogen atom, but is formulated on empirically sound concepts more closely linked with observable quantities — rather than the mysterious immaterial waves on which Schrödinger based his analysis. Unfortunately, any presentation of Heisenberg's theory makes heavy mathematical demands on the reader, so I will only mention that Schrödinger succeeded in deriving Heisenberg's approach from wave mechanics in another inspired paper in 1926 [37]. That being said, one cannot mention the name Heisenberg without mentioning his famous *uncertainty principle*. Born's probabilistic interpretation means that quantum mechanics studies the evolution of *probability measures* rather than the exact values of physical quantities. Heisenberg discovered that the laws of quantum theory place fundamental restrictions on the possible simultaneous probability distributions of such quantities, e.g., the distribution of the possible values of the position and velocity

INTRODUCTION

coordinates of a particle at a given instant of time. Whereas the *state* of a classical particle is characterised by its current position and velocity, Heisenberg's principle tells us that it is meaningless to discuss exact simultaneous values of position and velocity coordinates, only their statistical distributions, which are subject to the famous inequalities that bear Heisenberg's name. A proper discussion of this would take us well beyond the scope of this introduction and open up deeper questions about the meaning of quantum mechanics and its mathematical description. Quantum mechanics challenges us to rethink not only the dynamics of particles, but the mathematical theory of spaces on which it is founded. The adventurous reader is encouraged to follow this line of thought down the rabbit hole of current research, so let me end here by quoting one of the founding fathers of modern quantum theory, Paul Adrien Maurice Dirac, who noted in 1982 that [12]:

> *It seems clear that the present quantum mechanics is not in its final form. Some further changes will be needed, just about as drastic as the changes made in passing from Bohr's orbit theory to quantum mechanics.*

Let us now return to the story of atomic structure we left on page 16, and the discovery of a new particle that will prove to be the missing piece we need to finally unravel the puzzle of the atomic nucleus.

The Neutron

Although *protons* (or hydrogen nuclei) are not emitted spontaneously in the process of radioactive decay, in 1917-18 Ernest Rutherford managed to eject them artificially by bombarding the nuclei of nitrogen atoms with highly energetic alpha particles. Such an assault on heavier atoms had not succeed for the simple reason that their nuclear charge is too great, only the lightest atoms are vulnerable to attack because the alphas are too strongly repelled by the positive electric field surrounding heavier nuclei.[22]

The fact that Rutherford had managed to eject protons from the nucleus, together with the natural phenomena of β-decay, would seem to suggest that atomic nuclei do indeed consist of a dense amalgamation of protons and electrons as Russell recounts in his book. In the case of nitrogen, whose atomic weight is 14 and atomic number 7, the nucleus was thought to consist of 14 protons and seven 'nuclear electrons', to achieve the correct nuclear charge of $+7$ ($= 14 - 7$). However, this accounting did not fit with what the Italian physicist Franco Dino Rasetti[27, 28] observed in its spectrum. In its gaseous state, two nitrogen atoms combine to form the diatomic molecule N_2. Just like atoms, the rotational energy of diatomic molecules is *quantised*, and light of a definite frequency is emitted or absorbed when the molecule transitions from one state to another. Using Schrödinger's wave mechanics, one can show that half of the molecular energy levels observed in heteronuclear diatomic molecules[23] should be missing in homonuclear[24] diatomic molecules like N_2, which half depending on whether the nucleus consists of an odd or even number of constituents. Walter Heitler and Gerhard Herzberg[22] pointed out that Rasetti's results indicated that the nucleus of nitrogen should contain an even number of constituents, which was in stark contradiction with the assumption that it contains 7 'nuclear electrons' and 14 protons, $(14 + 7 = 21)$.

One way to resolve this dilemma was to abandon the idea that there were just two elementary particles, and introduce a third, a neutral particle with about the same mass as the proton. Rutherford had already toyed with the idea of a novel atomic nucleus called the *neutron*, with an atomic weight of about one and no electric charge, but he conceived of it as a bound state of a proton and a 'nuclear

---

[22]e.g., the nucleus of nitrogen carries a charge of just seven, whereas the nuclear charge of gold is 79, a more than 11 fold increase.

[23]i.e, molecules consisting of two different atoms, e.g., Carbon Monoxide – CO, Nitric Oxide – NO, etc.

[24]i.e., molecules consisting of identical atoms like N_2, O_2, etc.

INTRODUCTION

electron'. However, if the status of Rutherford's *neutron* was raised to that of an elementary particle, akin to that of the proton and electron, Rasetti's results are easily explained. Nitrogen nuclei could then be conceived as consisting of seven protons and seven *neutrons* — hence an even number of constituents — restoring harmony between theory and experiment. More generally, according to this idea the nuclei of an atom 'E' with atomic weight 'W' and atomic number 'Z', should consist of Z protons and $W - Z$ neutrons, e.g., the nucleus of a gold ($^{197}_{79}\text{Au}$) would comprise of 79 protons and 118 neutrons.

The discovery of such a neutral nuclear particle was made in 1932 by James Chadwick, one of Rutherford's former students. Chadwick was interested in some very confusing observations made by the husband and wife team Frédéric Joliot and Irène Curie[23]. When very fast alpha particles are directed at another light element, beryllium ($^{9}_{4}\text{Be}$), some "mysterious rays" are emitted that could not possibly be made up of protons. Nobody understood the nature of these rays, but they were far too deeply penetrating to consist of a charged particle like the proton, so the Joliot-Curies assumed they must be some form of electromagnetic radiation. However, when the couple directed the mysterious 'beryllium rays' at paraffin wax, they knocked out very fast protons. If the 'beryllium rays' were really electromagnetic in origin, Joliot and Curie realised that they could not account for the high energy of the ejected protons. It seemed energy was not conserved — more energy was being produced by the beryllium nuclei than imparted by the incident alpha particles. Chadwick soon resolved the paradox by studying how the rays affected other materials[6]. He measured how other light nuclei could be made to recoil by the rays, and noticed the correlation between the recoil velocity and atomic weight was exactly what one would expect if the beryllium rays were not electromagnetic, but made up of a neutral particle with mass very close to that of the proton[25]. Chadwick had discovered the *neutron*!

---

[25]The brilliant German physicist Werner Heisenberg was the first to speculate

ABC OF ATOMS

Energy

With the discovery of the neutron, you might think that our story is now complete. A cursory glance through any modern high school science book will turn up beautifully coloured pictures of atoms showing point like electrons orbiting a central core of *nucleons*[26] depicted as little coloured spheres. You now know that almost everything about this picture is wrong, yet it persists because of our desire to tell simple stories.

As we have seen, quantum mechanics, in its bizarre and mysterious way, reveals bound electrons as a kind of standing wave — *probability amplitudes* — in which the electron manifests itself not as a point particle in orbit, but as a kind of 'harmonic' representing a static *probability distribution*. It goes without mentioning that the inert little spheres making up the nucleus are also a fantasy that we may tell to young children — one that doesn't begin to do justice to the tremendous progress made in experimental and theoretical physics over the last 100 years. The protons and neutrons bound inside the nucleus must also be described in the language of quantum theory, but the forces that hold them together require a mathematical treatment far more sophisticated than anything even Russell could have dreamt of in 1923. Although we cannot describe that theory here, we can take a bird's-eye view of the whole subject in a simple and elegant way. There remains one ingredient in the recipe of atoms that we have only mentioned in passing, but one that is of the upmost importance to both theory and applications — the familiar yet abstract notion of *energy*. Unlike anything we have described so far, it is neither a particle or wave, nor anything we can even begin to draw. We cannot describe it in any specific form precisely because it's an aspect of *all* physical phenomena, somewhat akin to accounting in business. Richard Feynman described it thus[13]:

---

about the role of the neutron in the nucleus [18, 19, 20]. That the proton and neutron are so similar in every respect other than their electric charge, led to the idea that the nuclear force respects an *intrinsic symmetry* called *isospin*.

[26] the constituents of the nucleus, i.e, either protons or neutrons.

INTRODUCTION

It is important to realize that in physics today, we have no knowledge of what energy is. We do not have a picture that energy comes in little blobs of a definite amount. It is not that way. However, there are formulas for calculating some numerical quantity, and when we add it all together it gives "28" —always the same number. It is an abstract thing in that it does not tell us the mechanism or the reasons *for the various formulas.*

Yet *energy* is part of everyday life, and we all have an intuitive feeling for it despite its *administrative* nature. Physicists describe *energy* as the *potential to do work*; it's a measure of what is accomplished or could be accomplished when a force, any force no matter its character, acts in space and time. For example, everyone knows what a force is when they pick up a suitcase — the *weight* of the suitcase is the force of gravity[27] pulling down on your hand. Carrying that suitcase up a flight of stairs soon reveals the concept of energy — the more stairs you climb, the more energy you need. While your muscles become exhausted, the suitcase gains *potential energy*, and to balance the ledger it follows that your loss is the suitcase's gain. More precisely, the gravitational potential energy gained by the traveller[28] must exactly match the chemical energy used in climbing the stairs. In this way the accounting of energy reveals the cost of every transaction in the universe, and so confers a precise measure of any interaction provided we can quantify at least one side of the balance sheet. It follows that we may gain insights into the operation of forces for which we have neither direct experience nor theoretical understanding — and so it is we get a handle on the forces holding the nucleus together. But how can we possibly audit the energy transactions inside the nucleus of an atom? The answer is provided by Einstein's famous relation "$E = mc^2$", which states — quite miraculously — that this abstract bookkeeping phenomenon conjured up by physicists has that most material of characteristics,

---

[27] exerted by the mass of the entire planet.
[28] i.e., "suitcase" + "human".

mass. In particular when a nuclear interaction takes place, increases or decreases in the internal energy of participant nuclei must always be accompanied by a corresponding change in mass between the products and reactants according to Einstein's formula:

$$\Delta\ mass = \frac{\Delta\ internal\ energy}{speed\ of\ light^2}$$

A convenient scale for atomic mass is the *Dalton* (Da), which is defined as $\frac{1}{12}$ the mass of an unbound neutral atom of Carbon-12, i.e., the isotope $^{12}_{6}C$ taken at rest and in its nuclear and electronic ground state. With this measure, the mass of $^{12}_{6}C$ is exactly 12 Da by definition, but the mass of other atoms will not (in general) be a whole number of Daltons.

Figure 2 lists some familiar isotopes[29] along with their mass in Daltons. From this table it seems:

- lighter atoms tend to have values slightly greater than a whole number of Daltons (e.g., H, He, Be, N);

- middle weight atoms have values slightly less than a whole number of Daltons (e.g., K, Fe, Au);

- the heaviest elements again have values greater than whole numbers (e.g., Rn, Th, U).

A survey of all the known isotopes confirms this is indeed the case. Part of this variation is due to the difference in mass between protons and neutrons (neutrons are very slightly heavier than protons), but most of the variation is due to differences in the internal energy of nuclides[30]. That is to say the internal energy of the nucleus makes a greater contribution to the mass of atoms with low or high atomic number than it does to atoms in between. If follows that light nuclei

---

[29] Radon was called Niton prior to 1923.

[30] a distinct kind of atom or nucleus characterized by a specific number of protons and neutrons.

INTRODUCTION

Element	Isotope	Atomic mass (Da)
Hydrogen	$^{1}_{1}H$	1.007825
	$^{2}_{1}H$	2.0141
Helium	$^{4}_{2}He$	4.0026
Beryllium	$^{9}_{4}Be$	9.012183
Carbon	$^{12}_{6}C$	12
	$^{14}_{6}C$	14.003242
Nitrogen	$^{14}_{7}N$	14.003074
Oxygen	$^{16}_{8}O$	15.99491
	$^{17}_{8}O$	16.99910
Potassium	$^{39}_{19}K$	38.963706
	$^{40}_{19}K$	39.963998
Iron	$^{56}_{26}Fe$	55.934936
Gold	$^{197}_{79}Au$	196.966568
Radon	$^{222}_{86}Rn$	222.017577
Thorium	$^{232}_{90}Th$	232.03805
	$^{234}_{90}Th$	234.04360
Uranium	$^{235}_{92}U$	235.0439
	$^{238}_{92}U$	238.0508

Figure 2: Atomic mass of some isotopes.

can combine to release energy (*fusion*), and heavy nuclei can break apart to release energy (*radioactive decay, spontaneous fission*).

Fusion occurs naturally in the sun, but not on earth because it requires very extreme conditions to bring the nuclei close enough for the reaction to proceed[31]. As for radioactive decay, the only radioactive elements that remain in any abundance on earth are those that decay very slowly. For example $^{238}_{92}U$ undergoes α-decay to become $^{234}_{90}Th$:

$$^{238}_{92}U \longrightarrow {}^{234}_{90}Th + \alpha + \Delta \text{ internal energy}$$

---

[31] since all nuclei are positively charged and like charges repel.

ABC OF ATOMS

The difference in mass between the reactants and products is given by:

$$\Delta\ mass = \underbrace{238.0508}_{m_U} - (\underbrace{234.0436}_{m_{Th}} + \underbrace{4.0026}_{m_\alpha}) = 0.0046\ \text{Da}$$

This is $0.0046/238.0508$ of the mass of the parent $^{238}_{92}U$, and so by Einstein's formula the total energy per kilogram latent in $^{238}_{92}U$ is:

$$\frac{0.0046}{238.0508} \times (2.9979 \times 10^8\ \text{m/s})^2 = 1.7367 \times 10^{12}\ \text{J/kg}$$

The standard unit of energy revealed here is the *'joule'* (J), which corresponds to the *work done* or *energy expended* by a force of one *newton* (N)[32] acting over a distance of one metre (m). This is a prodigious amount of energy, but $^{238}_{92}U$ releases its energy very slowly. The half-life of $^{238}_{92}U$ is approximately 4.5×10^9 years, and so the α-decay channel takes $4\frac{1}{2}$ billion years to release just half of this amount of energy.

Of Atoms and Men

Spontaneous fission is a more exotic form of nuclear *disintegration* in which the nucleus of a heavy atom splits into various combinations of lighter nuclei. For example the nucleus of $^{238}_{92}U$ can also undergo spontaneous fission, but the process is very rare and occurs much more slowly than disintegration through α-decay[33]. Thus as far as humans are concerned, nuclear energy seemed to be a quite benign phenomenon until the morning of September 12, 1933. Following Adolf Hitler's rise to power, the Hungarian physicist Leo Szilard had moved to London and was living in the opulent luxury of the Imperial Hotel in Russell Square. Richard Rhodes describes the moment vividly in his book[29]:

---

[32] a Newton is the force which gives a mass of 1 kilogram an acceleration of 1 metre per second per second. The force of gravity near the earth's surface is approximately $9.8 N/kg$.

[33] the half-life of $^{238}_{92}U$ via spontaneous fission is estimated to be 8.4×10^{15} years.

INTRODUCTION

In London, where Southampton Row passes Russell Square, across from the British Museum in Bloomsbury, Leo Szilard waited irritably one gray Depression morning for the stoplight to change. A trace of rain had fallen during the night; Tuesday, September 12, 1933, dawned cool, humid and dull. Drizzling rain would begin again in early afternoon. When Szilard told the story later he never mentioned his destination that morning. He may have had none; he often walked to think. In any case another destination intervened. The stoplight changed to green. Szilard stepped off the curb. As he crossed the street time cracked open before him and he saw a way to the future, death into the world and all our woes, the shape of things to come...

Szilard had realized that if an element could be found in which neutrons would induce an energy producing nuclear reaction, and that amongst the products of that reaction there are more neutrons, then the process might cascade into a self-sustaining *"chain reaction"*. When neutron induced fission of Uranium was discovered by German scientists a few years later, Szilard immediately understood that an isotope of Uranium might be capable of sustaining such a chain reaction. We call such an isotope *fissile*.

Now, to be self-sustaining:

$$\begin{pmatrix} \text{\# neutrons} \\ \text{produced by fission} \end{pmatrix} \geq \begin{pmatrix} \text{\# neutrons absorbed in} \\ \text{the splitting of other nuclei} \end{pmatrix} + \begin{pmatrix} \text{\# neutrons that escape} \\ \text{into the surroundings} \end{pmatrix}$$

A *super critical mass* is an amount of fissile material sufficient to support an increasing rate of fission. Szilard could see that bringing together such a quantity of fissile material would furnish a bomb of explosive yield never before realized. A thought that would change his life, and also ignite the greatest arms race in human history.

Szilard went to work with Enrico Fermi, and on 2 December 1942 Fermi succeeded in inducing the first human-made self-sustaining nuclear chain reaction. The experiment was the first technical achievement of the Manhattan Project, a top secret effort initiated by Great

Britain, the United States and Canada to address concerns that Nazi Germany may be developing a nuclear weapon. The scientists realized that $^{235}_{92}$U is fissile, but the precise breakup of any single nucleus and the number of neutrons produced cannot be predicted, only their statistical spread. A typical example of a fission reaction is:

$$^{235}_{92}\text{U} + n \longrightarrow {}^{144}_{56}\text{Ba} + {}^{90}_{36}\text{Kr} + 2n + (\,3.2 \times 10^{-11}\text{J})$$

Now $1kg$ of $^{235}_{92}$U contains:

$$\frac{1000}{235} \times (\,\underbrace{6.022 \times 10^{23}}_{\text{Avogadro's number}}\,) \approx 4.26 \times (6.022 \times 10^{23})$$

$$\approx 2.6 \times 10^{24} \text{ atoms}$$

If each fission event releases an average of 2 neutrons, which both go on to split a further two $^{235}_{92}$U atoms, how many generations of doubling do we need to get to the order of 10^{24}? The answer is just 81 generations:

$$\sum_{t=0}^{80} 2^t = 2^{81} - 1 \approx 2.4 \times 10^{24}$$

Let's get an idea of the amount of energy released after 81 generations:

$$\text{Energy released} \approx 3.2 \times 10^{-11} J \times 2^{81}$$
$$\approx (32 \times 10^{-12} J) \times (2.4 \times 10^{24})$$
$$\approx 76.8 \times 10^{12} J$$

We can get a feeling for how much energy this represents by comparing it with the explosive yield of TNT (trinitrotoluene):

$$1kg \text{ of } {}^{235}_{92}\text{U releases} \approx \frac{76.8 \times 10^{12} J}{4 \times 10^6 J} \approx 19.2 kt \quad (1kg \text{ of TNT} \sim 4 \times 10^6 J)$$

INTRODUCTION

Here 'kt' refers to a unit of energy called the 'kiloton of TNT equivalent' or simply *kiloton*. One kiloton is defined to be exactly 4.184×10^9 joules, which is approximately the amount of energy released in the detonation of a metric ton (1000kg) of TNT. The yield of the bomb *Little Boy* dropped on Hiroshima was estimated to be about $15kt$, and so less than $1kg$ of the $64kg$ core of highly enriched uranium[34] underwent fission.

There is another way to make sense of such a large quantity of energy. The standard unit of energy — the joule — can also be understood as a *watt-second*. For example, 1 joule is the amount of energy consumed by a standard 100 watt incandescent bulb in $\frac{1}{100}^{th}$ of a second. The '*kilowatt-hour*' — the unit of energy familiar from your electricity bill — corresponds to $60 \times 60 \times 1000 = 3600000$ watt-seconds or joules. So the fission of $1kg$ of $^{235}_{92}$U releases

$$(76.8 \times 10^{12} J) \times 2.8 \times 10^{-7} kWh/J \approx 21.5 \times 10^6 kWh$$

i.e., 21.5 gigawatt-hours of energy. Here

$$1J = \frac{1}{3600} \times \frac{1}{1000} \approx 2.8 \times 10^{-7} kWh$$

The largest power station in the United Kingdom is the Drax power station in North Yorkshire. It has a maximum output capacity of 3.9 gigawatts, and so takes over $5\frac{1}{2}$ hours to generate the amount of energy released in the fission of $1kg$ of $^{235}_{92}$U. The dramatic difference is that a nuclear chain reaction can release this enormous quantity of energy in less than a microsecond!

Afterword

Understanding the atom has brought human civilization into a new era in which humans can harness the ultimate source of energy in the universe. The first atoms were formed about 380000 years after

---

[34] the highly enriched uranium in Little Boy was 80% $^{235}_{92}$U.

the creation of the universe. This primordial amorphous cosmos of hydrogen and helium was gradually shaped over billions of years into the stars and galaxies which make up our universe today. Most stars fade away after they exhaust their capacity to fuse hydrogen and helium in their cores. However, when the very largest stars in the universe — a tiny fraction of the 100 billion stars in a typical galaxy — run out of all options to fuse atoms together, they suddenly collapse under their own weight and explode, creating some of the most violent events in the universe. These *supernova* fuse enormous amounts of their gravitational energy into the formation of heavier elements and disperse their contents throughout interstellar space. You and I are made of that *star dust*, and as complex intelligent thinking beings we have now evolved to the point where we can harness the energy frozen into the cores of atoms. What we will do with this enormous potential is still uncertain. In the opening lines of the *Russell-Einstein Manifesto*, Russell wrote:

> *In the tragic situation which confronts humanity, we feel that scientists should assemble in conference to appraise the perils that have arisen as a result of the development of weapons of mass destruction, and to discuss a resolution in the spirit of the appended draft.*

Russell went on to organize that conference with his friend Joseph Rotblat, but he was unable to attend the meeting in person. The conference took place in July 1957 in Pugwash, Nova Scotia, the birth place of its financer, Cyrus Eaton. For the first time scientists from East and West came together to discuss highly sensitive issues. Rotblat and Russell believed "there was a more than even chance the meeting would break up in disagreement". Remarkably, the scientists found themselves in broad agreement on the main aspects of the agenda, and they decided to set up a new international organization, *Pugwash Conferences on Science and World Affairs*. Pugwash would go on to play a crucial role in ending the Cold War and, in 1995, the Nobel Peace Prize was awarded jointly to Rotblat and the Pugwash Conferences

INTRODUCTION

for their efforts to diminish the part played by nuclear arms in international politics and, in the longer run, to eliminate such arms.

Rotblat always gave credit to Russell as the founder of the Pugwash movement. Scientists are again being called upon to reduce the existential threat posed by nuclear weapons. Whether the world will listen to their pleas largely depends on the willingness of leaders as well as those they represent to learn *The ABC of Atoms*. I hope you will join us.

David Alexandre Ellwood
October 2023, London

Suggested Reading

I've listed here a personal selection of books for the reader who wishes to continue their journey into physics after finishing *The ABC of Atoms*:

- **Strange Glow: The Story of Radiation** by *Timothy J. Jorgensen*[2]. A very accessible account of the history and physics of radiation. Proof that the antidote to fear is better understanding.

- **Nucleus: A Trip into the Heart of Matter** by *Ray Mackintosh, Jim Al-Khalili, Bjorn Jonson, and Teresa Peña*[25]. A non-technical overview and great place to begin your study of nuclear physics.

- **Mr Tompkins in Paperback** by *George Gamow*[16]. Mr Tompkins is a bank teller who has incredible dreams which elucidate some of the most difficult concepts in modern physics. Brilliant and unforgettable. A great book with which to whet your appetite for more advanced physics.

- **Thirty Years that Shook Physics** by *George Gamow*[17]. A personal account of the historical development of quantum physics. Gamow was himself a leading theoretical physicist and knew all the key players in the field.

- **The Discovery of Subatomic Particles** by *Steven Weinberg*[40]. Weinberg takes you on a journey through history while teaching you a great deal of physics. I've relied on Weinberg's books since I was a student, and my account here draws closely on this book.

- **The First Three Minutes: A Modern View Of The Origin Of The Universe** by *Steven Weinberg*[41]. Weinberg's first popular science book and a classic. This book links the large with the small in physics, and teaches you a lot about everything in between.

- **QED: The Strange Theory of Light and Matter** by *Richard Feynman*[15]. A very unusual and brilliant book by a very unusual and brilliant physicist. This book takes you to the heart of what theoretical physicists do without relying on any of the advanced mathematics. It's a superb book to follow Russell's ABC.

- **The Infinity Puzzle: The Personalities, Politics, and Extraordinary Science Behind the Higgs Boson** by *Frank Close*[8]. This book brings the reader up to date with the latest advances in theoretical and experimental particle physics. Read it after reading the books by Weinberg and Feynman listed above.

- **The Making of the Atomic Bomb** by *Richard Rhodes*[29]. Simply the best history book on the subject.

- **Lectures on Quantum Mechanics** by *Steven Weinberg*[39]. Finally, for the serious student I must mention a serious textbook on quantum mechanics. If you decide to take your understanding to a professional level, learn from a master like Weinberg.

INTRODUCTION

References

- Felix Bloch. "Heisenberg and the early days of quantum mechanics". In: *Physics Today* 29.12 (Dec. 1976), pp. 23–27. ISSN: 0031-9228. DOI: 10.1063/1.3024633. eprint: https://pubs.aip.org/physicstoday/article-pdf/29/12/23/8280756/23_1_online.pdf.

- N. Bohr. "On the Constitution of Atoms and Molecules". In: *Philosophical Magazine*. 6th ser. 26 (1913), pp. 1, 476, 857.

- Max Born. "Quantenmechanik der Stoßvorgänge". In: *Zeitschrift für Physik* 38.11 (1926), pp. 803–827. DOI: 10.1007/BF01397184.

- Max Born. "Zur Quantenmechanik der Stoßvorgänge". In: *Zeitschrift für Physik* 37.12 (1926), pp. 863–867. DOI: 10.1007/BF01397477.

- Brian Cathcart. *Test of greatness : Britain's struggle for the atom bomb*. London: Murray London, 1994. ISBN: 0719552257; 9780719552250.

- J. Chadwick. "The Existence of a Neutron". In: *Proceedings of the Royal Society of London Series A* 136.830 (June 1932), pp. 692–708. DOI: 10.1098/rspa.1932.0112.

- W. Churchill. "Debate in the House of Commons; 3 November, 1953". In: *Commons Hansard* 520 (cc19–31).

- F. Close. *The Infinity Puzzle: The Personalities, Politics, and Extraordinary Science Behind the Higgs Boson*. OUP Oxford, 2013. ISBN: 9780199673308.

- C. Davisson and L. H. Germer. "Diffraction of Electrons by a Crystal of Nickel". In: *Physical Review* 30.6 (Dec. 1927), pp. 705–740. DOI: 10.1103/PhysRev.30.705.

- L. De Broglie. *Ondes et quanta*. Vol. 177. 1923, pp. 507, 548, 630.

- P.A.M. Dirac. *The Development of Quantum Theory: J. Robert Oppenheimer Memorial Prize Acceptance Speech*. J. Robert Oppenheimer. Gordon and Breach Science Publishers, 1971. ISBN: 9780677029702.

[12] P.A.M. Dirac. "The Early Years of Relativity". In: *The Centennial Symposium in Jerusalem*. Ed. by Gerald Holton and Yehuda Elkana. Princeton: Princeton University Press, 1982, pp. 79–90. ISBN: 9781400855438. DOI: 10.1515/9781400855438.79.

[13] R.P. Feynman, R.B. Leighton, and M. Sands. *The Feynman Lectures on Physics, Vol. I: The New Millennium Edition: Mainly Mechanics, Radiation, and Heat*. v. 1. Basic Books, 2015. ISBN: 9780465040858.

[14] R.P. Feynman, R.B. Leighton, and M. Sands. *The Feynman Lectures on Physics, Vol. III: The New Millennium Edition: Quantum Mechanics*. The Feynman Lectures on Physics. Basic Books, 2011. ISBN: 9780465025015.

[15] R.P. Feynman and A. Zee. *QED: The Strange Theory of Light and Matter*. Princeton Science Library. Princeton University Press, 2014. ISBN: 9781400847464.

[16] G. Gamow. *Mr Tompkins in Paperback*. Canto Classics. Cambridge University Press, 2012. ISBN: 9781107604681.

[17] G. Gamow. *Thirty Years that Shook Physics: The Story of Quantum Theory*. Dover Publications, 2012. ISBN: 9780486135168.

[18] W. Heisenberg. "Über den Bau der Atomkerne. I". In: *Zeitschrift für Physik* 77.1 (1932), pp. 1–11. DOI: 10.1007/BF01342433.

[19] W. Heisenberg. "Über den Bau der Atomkerne. II". In: *Zeitschrift für Physik* 78.3 (1932), pp. 156–164. DOI: 10.1007/BF01337585.

[20] W. Heisenberg. "Über den Bau der Atomkerne. III". In: *Zeitschrift für Physik* 80.9 (1933), pp. 587–596. DOI: 10.1007/BF01335696.

[21] W. Heisenberg. "Über quantentheoretische Umdeutung kinematischer und mechanischer Beziehungen." In: *Zeitschrift für Physik* 33.1 (1925), pp. 879–893. DOI: 10.1007/BF01328377.

[22] W. Heitler and G. Herzberg. "Gehorchen die Stickstoffkerne der Boseschen Statistik?" In: *Naturwissenschaften* 17.34 (1929), pp. 673–674. DOI: 10.1007/BF01506505.

INTRODUCTION

[23] Irène Joliot-Curie and Frédéric Joliot. "Émission de protons de grande vitesse par les substances hydrogénées sous l'influence des rayons γ très pénétrants". In: *Comptes Rendus des Séances de l'Académie des Sciences* 194 (1932), p. 273.

[24] T.J. Jorgensen. *Strange Glow: The Story of Radiation*. Princeton University Press, 2017. ISBN: 9780691178349.

[25] R. Mackintosh et al. *Nucleus: A Trip into the Heart of Matter*. Johns Hopkins University Press, 2011. ISBN: 9781421403519.

[26] V. Petkov, L. De Broglie, and A.J. Pomerans. *New Perspectives in Physics*. Minkowski Institute Press, 2021. ISBN: 9781989970607.

[27] F Rasetti. "On the Raman effect in diatomic gases". In: *Proceedings of the National Academy of Sciences* 15.3 (1929), pp. 234–237.

[28] Franco Rasetti. "On the raman effect in diatomic gases. ii". In: *Proceedings of the National Academy of Sciences* 15.6 (1929), pp. 515–519.

[29] R. Rhodes. *The Making of the Atomic Bomb*. Simon & Schuster, 2012. ISBN: 9781439126226.

[30] B. Russell. *Autobiography*. Routledge, 1998. ISBN: 9780415189859.

[31] B. Russell. "Debate in the House of Lords: "The International Situation"; 28 November, 1945". In: *Lords Hansard* 138 (cc88–93).

[32] E. Rutherford. "The Scattering of α and β Particles by Matter and the Structure of the Atom". In: *Philosophical Magazine*. 6th ser. 21 (1911), p. 669.

[33] E. Schrödinger. "Quantisierung als Eigenwertproblem". In: *Annalen Phys.* 384.4 (1926), pp. 361–376. DOI: 10.1002/andp.19263840404.

[34] E. Schrödinger. "Quantisierung als Eigenwertproblem". In: *Annalen Phys.* 384.6 (1926), pp. 489–527. DOI: 10.1002/andp.19263840602.

[35] E. Schrödinger. "Quantisierung als Eigenwertproblem". In: *Annalen Phys.* 385.13 (1926), pp. 437–490. DOI: 10.1002/andp.19263851302.

[36] E. Schrödinger. "Quantisierung als Eigenwertproblem". In: *Annalen Phys.* 386.18 (1926), pp. 109–139. DOI: 10.1002/andp.19263861802.

[37] Erwin Schrödinger. "Über das Verhältnis der Heisenberg-Born-Jordanschen Quantenmechanik zu der meinem". In: *Annalen der Physik* 384 (), pp. 734–756. URL: https://api.semanticschol.org/CorpusID:122923910.

[38] J.J. Thomson. "Cathode Rays". In: *Philosophical Magazine* 44 (1897), p. 295.

[39] S. Weinberg. *Lectures on Quantum Mechanics*. Cambridge University Press, 2015. ISBN: 9781107111660.

[40] S. Weinberg. *The Discovery of Subatomic Particles*. Scientific American library. Scientific American Books, 1983. ISBN: 9780716714880.

[41] S. Weinberg. *The First Three Minutes: A Modern View Of The Origin Of The Universe*. Basic Books, 2022. ISBN: 9781541603318.

CHAPTER I

INTRODUCTORY

To the eye or to the touch, ordinary matter appears to be continuous; our dinner-table, or the chairs on which we sit, seem to present an unbroken surface. We think that if there were too many holes the chairs would not be safe to sit on. Science, however, compels us to accept a quite different conception of what we are pleased to call " solid " matter; it is, in fact, something much more like the Irishman's definition of a net, " a number of holes tied together with pieces of string." Only it would be necessary to imagine the strings cut away until only the knots were left.

When science seeks to find the units of which matter is composed, it is led to continually smaller particles. The largest unit is the molecule, but a molecule is as a rule composed of " atoms " of several different " elements." For example, a molecule of water consists of two atoms of hydrogen and one of oxygen, which can be separated from each other by chemical methods. An atom, in its turn, is found to be a sort of solar system, with a sun and planets; the empty regions between the sun and the planets fill up vastly more space than they do, so that much the greater part of the volume that seems to us to be filled by a solid body is really unoccupied. In the solar system that constitutes an atom, the planets are called "electrons" and the sun is called the "nucleus." The nucleus itself is not simple except in the case of hydrogen; in all other cases, it is a complicated system consisting, in all likelihood, of electrons and hydrogen nuclei (or protons, as they are also called).

With electrons and hydrogen nuclei, so far as our present knowledge extends, the possibility of dividing up matter into bits comes to an end. No reason exists for supposing that these themselves have a structure, and are composed of still smaller bits. We do not know, of course, that reasons may not be found later for subdividing electrons and hydrogen nuclei; we only know that so far nothing prevents us from treating them as ultimate. It is difficult to know whether to be more astonished at the smallness of these

INTRODUCTORY

units, or at the fact that there are units, since we might have expected matter to be divisible *ad infinitum*. It will help us to picture the world of atoms if we have, to begin with some idea of the sizes of these units. Let us start with a gramme* of hydrogen, which is not a very large quantity. How many atoms will it contain? If the atoms were made up into bundles of a million million, and then we took a million million of these bundles, we should have about a gramme and a half of hydrogen. That is to say, the weight of one atom of hydrogen is about a million-millionth of a million-millionth of a gramme and a half. Other atoms weigh more than the atom of hydrogen, but not enormously more; an atom of oxygen weighs 16 times as much, an atom of lead rather more than 200 times as much. *Per contra*, an electron weighs very much less than a hydrogen atom; it takes about 1850 electrons to weigh as much as one hydrogen atom.

The space occupied by an atom is equally minute. As we shall see, an atom of a given kind is not always of the same size; when it is not crowded, the electrons which constitute its planets sometimes are much farther from

*A gramme is about one four-hundred-and-fifty-third of a pound.

its sun than they are under normal terrestrial conditions. But under normal conditions the diameter of a hydrogen atom is about a hundred-millionth of a centimetre (a centimetre is about a third of an inch). That is to say, this is about twice the usual distance of its one electron from the nucleus. The nucleus and the electron themselves are very much smaller than the whole atom, just as the sun and the planets are smaller than the whole region occupied by the solar system. The sizes of the electron and the nucleus are not accurately known, but they are supposed to be about a hundred thousand times as small as the whole atom.

It might be thought that not much could be known about such minute phenomena, since they are very far below what can be seen by the most powerful microscope. But in fact a great deal is known. What has been discovered about atoms by modern physicists is doubly amazing. In the first place, it is contrary to what every man of science expected, and in part very difficult to reconcile with other knowledge and with deep-seated prejudices. In the second place, it seems to the layman hardly credible that such very small things should be not only observable,

INTRODUCTORY

but measurable with a high degree of accuracy: Sherlock Holmes at his best did not show anything like the skill of the physicists in making inferences, subsequently verified, from minute facts which ordinary people would have thought unimportant. It is remarkable that, like Einstein's theory of gravitation, a great deal of the work on the structure of the atom was done during the war. It is probable that it will ultimately be used for making more deadly explosives and projectiles than any yet invented.

The study of the way in which atoms combine into molecules belongs to chemistry, and will not much concern us. We are concerned with the structure of atoms, the way in which electrons and nuclei come together to build up the various kinds of atoms. This study belongs to physics almost entirely. There are three methods by which most of our knowledge is obtained: the spectroscope, X-rays, and radio-activity. The hydrogen atom, which has a simple nucleus and only one electron is studied by means of the spectroscope almost alone. This is the easiest case, and the only one in which the mathematical difficulties can be solved completely. It is the case by means of which the most important principles were discovered and accurately tested. All the atoms except that of hydrogen present some problems which are too difficult for the mathematicians, in spite of the fact that they are largely of a kind that has been studied ever since the time of Newton. But although exact quantitative solutions of the questions that arise are often impossible, it is not impossible, even with the more complex atoms, to discover the *sort* of thing that is happening when they emit light or X-rays or radio-activity.

When an atom has many electrons, it seems that they are arranged in successive rings round the nucleus, all revolving round it approximately in circles or ellipses. (An ellipse is an oval curve, which may be described as a flattened-out circle.) The chemical properties of the atom depend, almost entirely, upon the outer ring; so does the light that it emits, which is studied by the spectroscope. The inner rings of electrons give rise to X-rays when they are disturbed, and it is chiefly by means of X-rays that their constitution is studied. The nucleus itself is the source of radio-activity. In radium and the other radio-active elements, the nucleus is unstable, and is apt to shoot out little particles with incredible velocity. As the nucleus is what really determines what sort of atom is

INTRODUCTORY

concerned, i.e. what element the atom belongs to, an atom which has ejected particles in radio-activity has changed its chemical nature, and is no longer the same element as it was before. Radio-activity has only been found among the heaviest atoms, which have the most complex structure. The fact that it occurs is one of the proofs that the nucleus of such elements has a structure and is complex. Until radio-activity was discovered, no process was known which changed one element into another. Now-a-days, transmutation, the dream of the alchemists, takes place in laboratories. But unfortunately it does not transform the baser metals into gold; it transforms radium, which is infinitely more valuable than gold, into lead—of a sort.

The simplest atom is that of hydrogen, which has a simple nucleus and a single electron. Even the one electron is lost when the atom is positively electrified: a positively electrified hydrogen atom consists of a hydrogen nucleus alone. The most complex atom known is that of uranium, which has, in its normal state, 92 electrons revolving round the nucleus, while the nucleus itself probably consists of 238 hydrogen nuclei and 146 electrons. No reason is known why there should not be still more complex atoms, and possibly such atoms may be discovered some day. But all the most complex atoms known are breaking down into simpler ones by radio-activity, so that one may guess that still more complex atoms could not be stable enough to exist in discoverable quantities.

The amount of energy packed up in an atom is amazing, considering its minuteness. There is least energy in the outer electrons, which are concerned in chemical processes, and yield, for instance, the energy derived from combustion. There is more in the inner electrons, which yield X-rays. But there is most in the nucleus itself. This energy in the nucleus only came to be known through radio-activity; it is the energy which is used up in the performances of radium. The nucleus of any atom except hydrogen is a tight little system, which may be compared to a family of energetic people engaged in a perpetual family quarrel. In radio-activity some members of the family emigrate, and it is found that the energy they used to spend on quarrels at home is sufficient to govern an empire. If this source of energy can be utilized commercially, it will probably in time supersede every other. Rutherford— to whom, more than any other single man,

INTRODUCTORY

is due the conception of the atom as a solar system of electrons revolving round a nucleus —is working on this subject, and investigating experimental methods of breaking up complex atoms into two or more simpler ones. This happens naturally in radio-activity, but only a few elements are radio-active, at any rate to an extent that we can discover. To establish the modern theory of the structure of nuclei on a firm basis, it is necessary to show, by artificial methods, that atoms which are not naturally radio-active can also be split up. For this purpose, Rutherford has subjected nitrogen atoms (and others) to a severe bombardment, and has succeeded in detaching hydrogen atoms from them. This whole investigation is as yet in its infancy. The outcome may in time revolutionize industry, but at present this is no more than a speculative possibility.

One of the most astonishing things about the processes that take place in atoms is that they seem to be liable to sudden discontinuities, sudden jumps from one state of continuous motion to another. The motion of an electron round its nucleus seems to be like that of a flea, which crawls for a while, and then hops. The crawls proceed accurately according to the old laws of dynamics, but the hops are a new phenomenon, concerning which certain totally new laws have been discovered empirically, without any possibility (so far as can be seen) of connecting them with the old laws. There is a possibility that the old laws, which represented motion as a smooth continuous process, may be only statistical averages, and that, when we come down to a sufficiently minute scale, everything really proceeds by jumps, like the cinema, which produces a misleading appearance of continuous motion by means of a succession of separate pictures.

In the following chapters, I shall try to explain in non-technical language what is known about the structure of atoms and how it has been discovered, in so far as this is possible without introducing any mathematical or other difficulties. Although a great deal is known, a great deal more is still unknown; at any moment, important new knowledge may be discovered. The subject is almost as interesting through the possibilities which it suggests as through what has actually been ascertained already; it is impossible to exaggerate the revolutionary effect which it may have both in the practice of industry and in the theory of physics.

CHAPTER II

THE PERIODIC LAW

Before we can understand the modern work on the structure of the atom, it is necessary to know something of the different kinds of atoms as they appear in chemistry. As every one knows, there are a great many different chemical " elements." The number known at present is eighty-seven, but new elements are discovered from time to time. The last discovery of a new element was announced as recently as January 22nd of this year (1923); this element was discovered in Copenhagen and has been christened hafnium. Each element consists of atoms of a special kind. As we saw in Chapter I, an atom is a kind of solar system, consisting of a nucleus which has electrons revolving round it. We shall see later that it is the nature of the nucleus that characterizes an element, and that two atoms of the same element may differ as to the number of their electrons and the shapes of their orbits. But for the present we are not concerned with the insides of atoms : we are taking them as units, in the way that chemistry takes them, and studying their outward behaviour.

The word "atom" originally meant "indivisible" and comes to us from the Greeks, some of whom believed that matter is composed of little particles which cannot be cut up. We know now that what are called atoms can be cut up, except in the case of positively electrified hydrogen (which consists of a hydrogen nucleus without any attendant electron). But in *chemistry*, apart from radio-activity, there is nothing to prove that atoms can be divided. So long as we could only study atoms by the methods of chemistry, that is to say, by their ways of combining with other atoms to form compounds, there was no way in which we could reach smaller units of matter out of which the atoms could be composed. Everything known before the discovery of radio-activity pointed to the view that an atom is indestructible, and this made it difficult to see how atoms could have a structure built out of smaller things, because, if they had, one would expect to find that the structure could be destroyed, just as a house

THE PERIODIC LAW

can be knocked down and reduced to a heap of bricks. We now know that in radio-activity this sort of thing does happen. Moreover it has proved possible, by means of the spectroscope, to discover with delicate precision all sorts of facts about the structure of the atom which were quite unknown until recent years.

It was of course recognized that science could not rest content with the theory that there were just eighty-seven different sorts of atoms. We could bring ourselves to believe that the universe is built out of two different sorts of things, or perhaps three ; we could believe that it is built out of an infinite number of different sorts of things. But some instinct rebels against the idea of its being built out of eighty-seven different sorts of things. The physicists have now all but succeeded in reducing matter to two different kinds of units, one (the proton or hydrogen nucleus) bearing positive electricity, and the other (the electron) bearing negative electricity. It is fairly certain that this reduction will prove to be right, but whether there is any further stage to be hoped for it is as yet impossible to say. What we can already say definitely is that the haphazard multiplicity of the chemical elements has given place to something more unified and systematic. The first step in this process, without which the later steps cannot be understood, was taken by the Russian chemist Mendeleeff, who discovered the " periodic law " of the elements.

The periodic law was discovered about the year 1870. At the time when it was discovered, the evidence for it was far less complete than it is at present. It has proved itself capable of predicting new elements which have subsequently been found and altogether the half-century that has passed since its discovery has enormously enhanced its importance. The elements can be arranged in a series by means of what is called their " atomic weight." By chemical methods, we can remove one element from a compound and replace it by an equal number of atoms of another element ; we can observe how much this alters the weight of the compound, and thus we can compare the weight of one kind of atom with the weight of another. The lightest atom is that of hydrogen ; the heaviest is that of uranium, which weighs over 238 times as much as that of hydrogen. It was found that, taking the weight of the hydrogen atom as one, the weights of a great many other atoms were almost exactly multiples of this unit, so that they were expressed by integers. The weight of the oxygen atom is a very little less than 16 times that of the hydrogen atom. It has been found convenient to *define* the atomic weight of oxygen as 16, so that the atomic weight of hydrogen becomes slightly more than one (1.008). The advantage of this definition is that it makes the atomic weights of a great many elements whole numbers, within the limits of accuracy that are possible in measurement. The recent work of F. W. Aston on what are called " isotopes " (concerning which we shall have more to say at a later stage) has shown that, in many cases where the atomic weight seems to be not a whole number, we really have a mixture of two different elements, each of which has a whole number for its atomic weight. This is what we should expect if the nuclei of the heavier atoms are composed of the nuclei of hydrogen atoms together with electrons (which are very much lighter than hydrogen nuclei). The fact that so many atomic weights are almost exactly whole numbers cannot be due to chance, and has long been regarded as a reason for supposing that atoms are built up out of smaller units.

Mendeleeff (and at about the same time the German chemist, Lothar Meyer) observed that an element would resemble in its properties, not those that came next to it in the series of atomic weights, but certain other elements which came at periodic intervals in the series. For example, there is a group of elements called "alkalis"; these are the 3rd, 11th, 19th, etc. in the series. These are all very similar in their chemical behaviour, and also in certain physical respects, notably their spectrum. Next to these come a group called "alkaline earths"; these are the 4th, 12th, 20th, etc. in the series. The third group are called "earths." There are eight such groups in all. The eighth, which was not known when the law was discovered, is the very interesting group of "inert gases," Helium, Neon, Argon, Krypton, Xenon, and Niton, all discovered since the time of Mendeleeff. These are the 2nd, 10th, 18th, 36th, 54th and 86th respectively in the series of elements. They all have the property that they will not enter into chemical combinations with any other elements; the Germans, on this account, call them the "noble" gases. The elements from an alkali to the next inert gas form what is called one "period." There are seven periods altogether.

THE PERIODIC LAW

When once the periodic law had been discovered, it was found that a great many properties of elements were periodic. This gave a principle of arrangement of the elements, which in the immense majority of cases placed them in the order of their atomic weights, but in a few cases reversed this order on account of other properties. For example, argon, which is an inert gas, has the atomic weight 39.88, whereas potassium, which is an alkali, has the smaller atomic weight 39.10 Accordingly argon, in spite of its greater atomic weight, has to be placed before potassium, at the end of the third period, while potassium has to be put at the beginning of the fourth. It has been found that, when the order derived from the periodic law differs from that derived from the atomic weight, the order derived from the periodic law is much more important; consequently this order is always adopted.

When the periodic law was first discovered there were a great many gaps in the series that is to say, the law indicated that there ought to be an element with such-and-such properties at a certain point in the series, but no such element was known. Confidence in the law was greatly strengthened by the discovery of new elements having the requisite properties. There are now only five gaps remaining.

The seven periods are of very unequal length. The first contains only two elements, hydrogen and helium. The second and third each contain eight; the fourth contains eighteen, the fifth again contains eighteen, the sixth thirty-two, and the seventh only six. But the seventh, which consists of radio-active elements, is incomplete; its later members would presumably be unstable, and break down by radio-activity. Niels Bohr* suggests that, if it were complete, it would again contain thirty-two elements, like the sixth period.

By means of the periodic law, the elements are placed in a series, beginning with hydrogen and ending with uranium. Counting the four gaps, there are ninety-two places in the series. What is called the " atomic number " of an element is simply its place in this series. Thus hydrogen has the atomic number 1, and uranium has the atomic number 92. Helium is 2, Lithium is 3, carbon 6, nitrogen 7, oxygen 8, and so on. Radium, which fits quite correctly into the series, is 88. The atomic number is

* *The Theory of Spectra and Atomic Constitution*
Cambridge, 1922, pp. 112–3.

THE PERIODIC LAW

much more important than the atomic weight; we shall find that it has a very simple interpretation in the structure of the atom.

It has lately been discovered that there are sometimes two or more slightly different elements having the same atomic number. Such elements are exactly alike in their chemical properties, their optical spectra, and even their X-ray spectra; they differ in no observable property except their atomic weight. It is owing to their extreme similarity that they were not distinguished sooner. Two elements which have the same atomic number are called "isotopes." We shall return to them when we come to the subject of radio-activity, when it will appear that their existence ought not to surprise us. For the present we shall ignore them, and regard as identical two elements having the same atomic number.

There are irregularities in the periodicity of the elements, which we are only now beginning to understand. The second and third periods, which each contain eight elements, are quite regular; the first element in the one is like the first in the other, the second like the second, and so on. But the fourth period has 18 elements, so that its elements cannot correspond one by one to those of the third period.

There are eight elements with new properties (the 21st to the 28th), and others in which the correspondence is not exact. The fifth period corresponds regularly, element for element, with the fourth, which is possible because both contain 18 elements. But in the sixth period there are 32 elements, and 16 of these, including one unknown (the "rare earths"), do not correspond to any of the elements in earlier periods. Niels Bohr, in the book mentioned above, has offered ingenious explanations of these apparent irregularities, which are still more or less hypothetical, but are probably in the main correct. Some very important facts, however, remain quite unexplained, notably the fact that iron and the two neighbouring elements have magnetic properties which are different in a remarkable way from those of all other elements.

The atomic weight of the earlier elements (except hydrogen) is double, or one more than double, the atomic number. Thus helium, the second element, has the atomic weight 4; lithium, the third, has the atomic weight 7 (very nearly); oxygen, the eighth, has the atomic weight 16. But after the 20th element the atomic weight becomes increasingly more than double the atomic number. For instance, silver, the 47th element, has atomic weight 107.88 ; gold, the 79th, has atomic weight 197.2 ; uranium, the 92nd, has atomic weight 238.2.

It is remarkable that X-ray spectra, which were unknown until a few years ago, show a perfectly regular progression throughout the whole series of elements, even in those cases where the order of the periodic table departs from the order of the atomic weights. This is a striking confirmation of the correctness of the order that has been adopted.

The fact of the periodic relations among the elements, and of progressive properties such as those shown in X-ray spectra (which we shall consider later on), is enough to make it highly probable that there are relations between different kinds of atoms, of a sort which implies that they are all built out of common materials, which must be regarded as the true " atoms " in the philosophical sense, i.e. the indivisible constituents of all matter. Chemical atoms are not indivisible, but are composed of simpler constituents which are indivisible, so far as our present knowledge goes. Without the knowledge of the periodic law, it is probable that the modern theories of the constitution of atoms would never have been discovered ; *per contra*, the facts embodied in the periodic law form an essential part of the basis for these theories. The broad lines of atomic constitution will be explained in the next chapter.

CHAPTER III

ELECTRONS AND NUCLEI

An atom, as we saw in Chapter I, consists, like the solar system, of a number of planets moving round a central body, the planets being called " electrons " and the central body a " nucleus." But the planets are not attached as firmly to the central body as they are in the solar system. Sometimes, under outside influences, a planet flies off, and either becomes attached to some other system, or wanders about for a while as a free electron. Under certain circumstances, the path of a free electron can actually be photographed; so can the paths of helium nuclei that are momentarily destitute of attendant electrons. This is done by making them travel through water vapour, which enables each to collect a little cloud, and so become large enough to be visible with a powerful microscope. These observations of individual electrons and helium nuclei are extraordinarily instructive.

They travel most of their journey in nearly straight lines, but are liable to sudden deviations when they find themselves very near to the electrons or nuclei of atoms that stand in their way. Helium nuclei are much less easily deflected from the straight line than electrons, showing that they have much greater mass. By exposing these particles to electric and magnetic forces and observing the effect upon their motion, it is possible to calculate their velocity and their mass. By one means or another, it is possible to find out just as much about them as we can find out about larger bodies.

An atom differs from the solar system by the fact that it is not gravitation that makes the electrons go round the nucleus, but electricity. As everybody knows, there are two kinds of electricity, positive and negative These are mere names; the two kinds might just as well be called " A " and " B." None of the ideas commonly associated with the words " positive " and " negative " must be allowed to intrude when we speak of positive and negative electricity. Each kind of electricity attracts its opposite and repels its own kind, like male and female. It is very easy to see electrical attraction in operation. For instance, take a piece of sealing-wax and rub it for a while on your sleeve. You will find that it will pick up small bits of paper if it is held a little distance above them, just as a magnet will pick up a needle. The sealing-wax attracts the bits of paper because it has become electrified by friction. In a similar way, the central nucleus of an atom, which consists of positive electricity, attracts the electrons, which consist of negative electricity. The law of attraction is the same as in the solar system : the nearer the nucleus and the electron are to each other the greater is the attraction, and the attraction increases faster than the distance diminishes. At half the distance, there is four times the attraction ; at a third of the distance, nine times ; at a quarter, sixteen times, and so on. This is what is called the law of the inverse square. But whereas the planets of the solar system attract one another, the electrons in an atom, since they all have negative electricity, repel one another, again according to the law of the inverse square.

Some readers may expect me at this stage to tell them what electricity " really is." The fact is that I have already said what it is. It is not a thing, like St. Paul's Cathedral ; it is a way in which things behave. When we have told how things behave when they are electrified, and under what circumstances they are electrified, we have told all there is to tell. When we are speaking of large bodies, there are three states possible : they may be more or less positively electrified, or more or less negatively electrified, or neutral. Ordinary bodies at ordinary times are neutral, but in a thunderstorm the earth and the clouds have opposite kinds of electricity. Ordinary bodies are neutral because their small parts contain equal amounts of positive and negative electricity ; the smallest parts, the electrons and nuclei, are never neutral, the electrons always having negative electricity and the nuclei always having positive electricity. That means simply that electrons repel electrons, nuclei repel nuclei, and nuclei attract electrons, according to certain laws ; that they behave in a certain way in a magnetic field ; and so on. When we have enumerated these laws of behaviour, there is nothing more to be said about electricity, unless we can discover further laws, or simplify and unify the statement of the laws already known. When I say that an electron has a certain amount of negative electricity, I mean merely that it

ELECTRONS AND NUCLEI

behaves in a certain way. Electricity is not like red paint, a substance which can be put on to the electron and taken off again; it is merely a convenient name for certain physical laws.

All electrons, whatever kind of atom they may belong to, and also if they are not attached to any atom, are exactly alike—so far, at least, as the most delicate tests can discover. Any two electrons have exactly the same amount of negative electricity, the smallest amount that can exist. They all have the same mass. (This is not the mass directly obtained from measurement, because when an electron is moving very fast its measured mass increases. This is true, not only of electrons, but of all bodies, for reasons explained by the theory of relativity, which will concern us at a later stage. But with ordinary bodies the effect is inappreciable, because they all move very much more slowly than light. Electrons, on the contrary, have been sometimes observed to move with velocities not much less than that of light; they can even reach 99 per cent. of the velocity of light. At this speed, the increase of measured mass is very great. But when we introduce the correction demanded by the theory of relativity, it is found that the mass of any two electrons is the same.) Electrons also all have the same size, in so far as they can be said to have a definite size. (For reasons which will appear later, the notion of the " size " of an electron is not so definite as we should be inclined to think.) They are the ultimate constituents of negative electricity, and one of the two kinds of ultimate constituents of matter.

Nuclei, on the contrary, are different for different kinds of elements. We will begin with hydrogen, which is the simplest element. The nucleus of the hydrogen atom has an amount of positive electricity exactly equal to the amount of negative electricity on an electron. It has, however, a great deal more ordinary mass (or weight) ; in fact, it is about 1850 times as heavy as an electron, so that practically all the weight of the atom is due to the nucleus. When positive and negative electricity are present in equal amounts in a body, they neutralize each other from the point of view of the outside world, and the body appears as unelectrified. When a body appears as electrified, that is because there is a preponderance of one kind of electricity. The hydrogen atom, when it is unelectrified,

ELECTRONS AND NUCLEI 35

consists simply of a hydrogen nucleus with one electron. If it loses its electron, it becomes positively electrified. Most kinds of atoms are capable of various degrees of positive electrification, but the hydrogen atom is only capable of one perfectly definite amount. This is part of the evidence for the view that it has only one electron in its neutral condition. If it had two in its neutral condition, the amount of positive electricity in the nucleus would have to be equal to the amount of negative electricity in two electrons, and the hydrogen atom could acquire a double charge of positive electricity by losing both its electrons. This sometimes happens with the helium atom, and with the heavier atoms, but never with the hydrogen atom.

Under normal conditions, when the hydrogen atom is unelectrified, the electron simply continues to go round and round the nucleus, just as the earth continues to go round and round the sun. The electron may move in any one of a certain set of orbits, some larger, some smaller, some circular, some elliptical. (We shall consider these different orbits presently.) But when the atom is undisturbed, it has a preference for the smallest of the circular orbits, in which, as we saw in

Chapter I, the distance between the nucleus and the electron is about half a hundred-millionth of a centimetre. It goes round in this tiny orbit with very great rapidity; in fact its velocity is about one hundred-and-thirty-fourth of the velocity of light, which is 300,000 kilometres (about 180,000 miles) a second. Thus the electron manages to cover about 2,200 kilometres (or about 1400 miles) in every second. To do this, it has to go round its tiny orbit about seven thousand million times in a millionth of a second; that is to say, in a millionth of a second it has to live through about seven thousand million of its "years." The modern man is supposed to have a passion for rapid motion, but nature far surpasses him in this respect.

It is odd that, although the hydrogen nucleus is very much heavier than an electron, it is probably no larger. The dimensions of an electron are estimated at about a hundred thousandth of the dimensions of its orbit. This, however, is not to be taken as a statement with a high degree of accuracy; it merely gives the sort of size that we are to think of. As for the nucleus, we know that in the case of hydrogen, it is probably about the same size as an electron, but we do not

ELECTRONS AND NUCLEI

know this for certain. The hydrogen nucleus may be quite without structure, like an electron, but the nuclei of other elements have a structure, and are probably built up out of hydrogen nuclei and electrons.

As we pass up the periodic series of the elements, the positive charge of electricity in the nucleus increases by one unit with each step. Helium, the second element in the table, has exactly twice as much positive electricity in its nucleus as there is in the nucleus of hydrogen; lithium, the third element, has three times as much; oxygen, the eighth, has eight times as much; uranium, the ninety-second (counting the gaps), has ninety-two times as much. Corresponding to this increase in the positive electricity of the nucleus, the atom in its unelectrified state has more electrons revolving round the nucleus. Helium has two electrons, lithium three, and so on, until we come to uranium, like the Grand Turk, with ninety-two consorts revolving round him. In this way the negative electricity of the electrons exactly balances the positive electricity of the nucleus, and the atom as a whole is electrically neutral. When, by any means, an atom is robbed of one of its electrons, it becomes positively electrified;

if it is robbed of two electrons, it becomes doubly electrified and remains electrified until it has an opportunity of annexing from elsewhere as many electrons as it has lost. A body can be negatively electrified by containing free electrons; an atom may for a short time have more than its proper number of electrons, and thus become negatively electrified, but this is an unstable condition, except in chemical combinations.

Nobody knows exactly how the electrons are arranged in other atoms than hydrogen. Even with helium, which has only two electrons, the mathematical problems are too difficult to be solved completely; and when we come to atoms that have a multitude of electrons, we are reduced largely to guesswork. But there is reason to think that the electrons are arranged more or less in rings, the inner rings being nearer to the nucleus than the outer ones. We know that the electrons must all revolve about the nucleus in orbits which are roughly circles or ellipses, but they will be perturbed from the circular or elliptic path by the repulsions of the other electrons. In the solar system, the attractions which the planets exercise upon each other are very minute compared to the attraction of the

ELECTRONS AND NUCLEI

sun, so that each planet moves very nearly as if there were no other planets. But the electrical forces between two electrons are not very much less strong than the forces between electrons and nucleus at the same distance. In the case of helium, they are half as strong; with lithium, a third as strong, and so on. This makes the perturbations much greater than they are in the solar system, and the mathematics correspondingly more difficult. Moreover we cannot actually observe the orbits of the electrons, as we can those of the planets; we can only infer them by calculations based upon data derived mainly from the spectrum of the element concerned, including the X-ray spectrum.

We shall have more to say at a later stage about the nature of these rings, which cannot be as simple as was supposed at first. The first hypothesis was that the electrons were like the people in a merry-go-round, all going round in circles, some in a small circle near the centre, others in a larger circle, others in a still larger one. But for various reasons the arrangement cannot be as simple as that. In spite of uncertainties of detail, however, it remains practically certain that there are successive rings of electrons; one ring in atoms belonging to the first period, two in the second period, three in the third, and so on. Each period begins with an alkali, which has only one electron in the outermost ring, and ends with an inert gas, which has as many electrons in the outermost ring as it can hold. It is impossible to get a ring to hold more than a certain number of electrons, though it has been suggested by Niels Bohr, in an extremely ingenious speculation, that a ring can hold more electrons when it has other rings outside it than when it is the outer ring. His theory accounts extraordinarily well for the peculiarities of the periodic table, and is therefore worth understanding, though it cannot yet be regarded as certainly true.

The previous view was that each ring, when complete, held as many electrons as there are elements in the corresponding period. Thus the first period contains only two elements (hydrogen and helium); therefore the innermost ring, which is completed in the helium atom, must contain two electrons. This remains true on Bohr's theory. The second period consists of eight elements, and is completed when we reach neon. The unelectrified atom of neon, therefore, will have two electrons in the inner ring and eight in the outer. The third period again consists of eight elements, ending with argon; therefore argon, in its neutral state, will have a third ring consisting of a further eight electrons. So far, we have not reached the parts of the periodic table in which there are irregularities, and therefore Bohr accepts the current view, except for certain refinements which need not concern us at present. But in the fourth period, which consists of 18 elements, there are a number of elements which do not correspond to earlier ones in their chemical and spectroscopic properties. Bohr accounts for this by supposing that the new electrons are not all in the new outermost ring, but are some of them in the third ring, which is able to hold more when it has other electrons outside it. Thus krypton, the inert gas which completes the fourth period, will still have only eight electrons in its outer ring, but will have eighteen in the third ring. Some elements in the fourth period differ from their immediate predecessors, not as regards the outer ring, but by having one more electron in the third ring. These are the elements that do not correspond accurately to any elements in the third period. Similarly the fifth period, which again consists of 18 elements, will add its new electrons partly in the new fifth ring, partly in the fourth, ending with Xenon, which will have eight electrons in the fifth ring, eighteen in the fourth, and the other rings as in krypton. In the sixth period, which has 32 elements, the new electrons are added partly in the sixth ring, partly in the fifth, and partly in the fourth; the rare earths are the elements which add new electrons in the fourth ring. Niton, the inert gas which ends the sixth period, will, according to Bohr's theory, have its first three rings the same as those of Krypton and Xenon, but its fourth ring will have 32 electrons, its fifth 18, and its sixth eight. This theory has very interesting niceties, which, however, cannot be explained at our present stage.

The chemical properties of an element depend almost entirely upon the outer ring of electrons, and that is why they are periodic. If we accept Bohr's theory, the outer ring, when it is completed, always has eight electrons, except in hydrogen and helium. There is a tendency for atoms to combine so as to make up the full number of electrons in the outer ring. Thus an alkali, which has one electron in the outer ring, will combine readily with an element that comes just before an

ELECTRONS AND NUCLEI

inert gas, and so has one less electron in the outer ring than it can hold. An element which has two electrons in the outer ring will combine with an element next but one before an inert gas, and so on. Two atoms of an alkali will combine with one atom of an element next but one before an inert gas. An inert gas, which has its outer ring already complete, will not combine with anything.

CHAPTER IV

THE HYDROGEN SPECTRUM

The general lines of atomic structure which have been sketched in previous chapters have resulted largely from the study of radio-activity, together with the theory of X-rays and the facts of chemistry. The general picture of the atom as a solar system of electrons revolving about a nucleus of positive electricity is derived from a mass of evidence, the interpretation of which is largely due to Rutherford ; to him also is due a great deal of our knowledge of radio-activity and of the structure of nuclei. But the most surprising and intimate secrets of the atom have been discovered by means of the spectroscope. That is to say, the spectroscope has supplied the experimental facts, but the interpretation of the facts required an extraordinarily brilliant piece of theorizing by a young Dane, Niels Bohr, who, when he first propounded his theory (1913), was still working under Rutherford.

THE HYDROGEN SPECTRUM

The original theory has since been modified and developed, notably by Sommerfeld, but everything that has been done since has been built upon the work of Bohr. This chapter and the next will be concerned with his theory in its original and simplest form.

When the light of the sun is made to pass through a prism, it becomes separated by refraction into the different colours of the rainbow. The spectroscope is an instrument for effecting this separation into different colours for sunlight or for any other light that passes through it. The separated colours are called a spectrum, so that a spectroscope is an instrument for seeing a spectrum. The essential feature of the spectroscope is the prism through which the light passes, which refracts different colours differently, and so makes them separately visible. The rainbow is a natural spectrum, caused by refraction of sunlight in raindrops.

When a gas is made to glow, it is found by means of the spectroscope that the light which it emits may be of two sorts. The way in which these are optically distinguished is not important for our purposes: what matters to us is the difference in the way they are caused. The first sort, which are called "band-spectra," are due to molecules; the second sort, called "line-spectra," are due to atoms. The first sort will not further concern us; it is from line-spectra that our knowledge of atomic constitution is obtained.

When white light is passed through a gas that is not glowing, and then analysed by the spectroscope, it is found that there are dark lines, which are to a great extent (though not by any means completely) identical with the bright lines that were emitted by the glowing gas. These dark lines are called the "absorption-spectrum" of the gas, whereas the bright lines are called the "emission-spectrum."

Every element has its characteristic spectrum, by which its presence may be detected. The spectrum, as we shall see, depends in the main upon the electrons in the outer ring. When an atom is positively electrified by being robbed of an electron in the outer ring, its spectrum is changed, and becomes very similar to that of the preceding element in the periodic table. Thus positively electrified helium has a spectrum closely similar to that of hydrogen—so similar that for a long time it was mistaken for that of hydrogen.

The spectra of elements known in the laboratory are found in the sun and the stars,

THE HYDROGEN SPECTRUM 47

thus enabling us to know a great deal about the chemical constitution of even the most distant fixed stars. This was the first great discovery made by means of the spectroscope.

The application of the spectroscope that concerns us is different. We are concerned with the explanation of the lines emitted by different elements. Why does an element have a spectrum consisting of certain sharp lines ? What connection is there between the different lines in a single spectrum ? Why are the lines sharp instead of being diffuse bands of colours? Until recent years, no answer whatever was known to these questions ; now the answer is known with a considerable approach to completeness. In the two cases of hydrogen and positively electrified helium, the answer is exhaustive ; everything has been explained, down to the tiniest peculiarities. It is quite clear that the same principles that have been successful in these two cases are applicable throughout, and in part the principles have been shown to yield observed results ; but the mathematics involved in the case of atoms that have many electrons is too difficult to enable us to deduce their spectra completely from theory, as we can in the simplest cases. In the cases that can be worked out, the calculations are not difficult. Those who are not afraid of a little mathematics can find an outline in Norman Campbell's "Series Spectra" (Cambridge, 1921), and a fuller account in Sommerfeld's "Atomic Structure and Spectral Lines," of which an English translation has lately been published by Methuen.

As every one knows, light consist of waves. Light-waves are distinguished from sound-waves by being what is called "transverse," whereas sound-waves are what is called "longitudinal." It is easy to explain the difference by an illustration. Suppose a procession marching up Piccadilly. From time to time the police will make them halt in Piccadilly Circus; whenever this happens, the people behind will press up until they too have to halt, and a wave of stoppage will travel all down the procession. When the people in front begin to move on, they will thin out, and the process of thinning out will travel down the whole procession just as the previous process of condensation did. This is what a sound-wave is like; it is called a "longitudinal" wave, because the people move all the time in the same direction in which the wave moves. But now suppose a mounted

THE HYDROGEN SPECTRUM 49

policeman, whose duty it is to keep half the road clear, rides along the right-hand edge of the procession. As he approaches, the people on the right will move to the left, and this movement to the left will travel along the procession as the policeman rides on. This is a " transverse " wave, because, while the wave travels straight on, the people move from right to left, at right angles to the direction in which the wave is travelling. This is the way a light-wave is constructed; the vibration which makes the wave is at right angles to the direction in which the wave is travelling.

This is, of course, not the only difference between light-waves and sound-waves. Sound waves only travel about a mile in five seconds, whereas light-waves travel about 180,000 miles a second. Sound-waves consist of vibrations of the air, or of whatever material medium is transmitting them, and cannot be propagated in a vacuum; whereas light-waves require no material medium. People have invented a medium, the aether, for the express purpose of transmitting light-waves. But all we really know is that the waves are transmitted; the aether is purely hypothetical, and does not really add anything to our knowledge. We know the mathematical properties of light-waves, and the sensations they produce when they reach the human eye, but we do not know what it is that undulates. We only suppose that something must undulate because we find it difficult to imagine waves otherwise.

Different colours of the rainbow have different wave-lengths, that is to say, different distances between the crest of one wave and the crest of the next. Of the visible colours, red has the greatest wave-length and violet the smallest. But there are longer and shorter waves, just like those that make light, except that our eyes are not adapted for seeing them. The longest waves of this sort that we know of are those used in wireless-telegraphy, which sometimes have a wave-length of several miles. X-rays are rays of the same sort as those that make visible light, but very much shorter; γ-rays, which occur in radio-activity, are still shorter, and are the shortest we know. Many waves that are too long or too short to be seen can nevertheless be photographed. In speaking of the spectrum of an element, we do not confine ourselves to visible colours, but include all experimentally discoverable waves of the same sort as those that make visible colours. The X-ray spectra, which are in

THE HYDROGEN SPECTRUM 51

some ways peculiarly instructive, require quite special methods, and are a recent discovery, beginning in 1912. Between the wave-lengths of wireless-telegraphy and those of visible light there is a vast gap; the wave-lengths of ordinary light (including ultra-violet) are between a ten-thousandth and about a hundred-thousandth of a centimetre. There is another long gap between visible light and X-rays, which are on the average composed of waves about ten thousand times shorter than those that make visible light. Between X-rays and γ-rays there is no gap.

In studying the connection between the different lines in the spectrum of an element, it is convenient to characterize a wave, not by its wave-length, but by its "wave-number," which means the number of waves in a centimetre. Thus if the wave-length is one ten-thousandth of a centimetre, the wave-number is 10,000; if the wave-length is one hundred-thousandth of a centimetre, the wave-number is 100,000, and so on. The shorter the wave-length, the greater is the wave-number. The laws of the spectrum are simpler when they are stated in terms of wave-numbers than when they are stated in terms of wave-lengths. The wave-number is also sometimes called the "frequency," but this term is more properly employed to express the number of waves that pass a given place in a second. This is obtained by multiplying the wave-number by the number of centimetres that light travels in a second, i.e. thirty thousand million. These three terms, wave-length, wave-number, and frequency must be borne in mind in reading spectroscopic work.

In stating the laws which determine the spectrum of an element, we shall for the present confine ourselves to hydrogen, because for all other elements the laws are less simple.

For many years no progress was made towards finding any connection between the different lines in the spectrum of hydrogen. It was supposed that there must be one fundamental line, and that the others must be like harmonics in music. The atom was supposed to be in a state of complicated vibration, which sent out light-waves having the same frequencies that it had itself. Along these lines, however, the relations between the different lines remained quite undiscoverable.

At last, in 1908, a curious discovery was made by W. Ritz, which he called the Principle of Combination. He found that all the lines were connected with a certain number of

THE HYDROGEN SPECTRUM 53

inferred wave-numbers which are called " terms," in such a way that every line has a wave-number which is the difference of two terms, and the difference between any two terms (apart from certain easily explicable exceptions) gives a line. The point of this law will become clearer by the help of an imaginary analogy. Suppose a shop belonging to an eccentric shopkeeper had gone bankrupt, and it was your business to look through the accounts. Suppose you found that the only sums ever spent by customers in the shop were the following: 19/11, 19/-, 15/-, 10/-, 9/11, 9/-, 5/-, 4/11, 4/-, 11d. At first these sums might seem to have no connection with each other, but if it were worth your while you might presently notice that they were the sums that would be spent by customers who gave 20/-, 10/-, 5/- or 1/- and got 10/-, 5/- 1/-, or 1d. in change. You would certainly think this very odd, but the oddity would be explained if you found that the shopkeeper's eccentricity took the form of insisting upon giving one coin or note in change, no more and no less. The sums spent in the shop correspond to the lines in the spectrum, while the sums of 20/-, 10/-, 5/-, 1/-, and 1d. correspond to the terms. You will observe that there are more lines than terms (10 lines and 5 terms, in our illustration). As the number of both increases, the disproportion grows greater; 6 terms would give 15 lines, 7 terms would give 21, 8 would give 28, 100 would give 4950. This shows that, the more lines and terms there are, the more surprising it becomes that the Principle of Combination should be true, and the less possible it becomes to attribute its truth to chance. The number of lines in the spectrum of hydrogen is very large.

The terms of the hydrogen spectrum can all be expressed very simply. There is a certain fundamental wave-number, called Rydberg's constant after its discoverer. Rydberg discovered that this constant was always occurring in formulae for series of spectral lines, and it has been found that it is very nearly the same for all elements. Its value is about 109700 waves per centimetre. This may be taken as the fundamental term in the hydrogen spectrum. The others are obtained from it by dividing it by 4 (twice two), 9 (three times three), 16 (four times four), and so on. This gives all the terms; the lines are obtained by subtracting one term from another. Theoretically, this rule gives an infinite number of terms, and therefore of lines; but in practice

THE HYDROGEN SPECTRUM 55

the lines grow fainter as higher terms are involved, and also so close together that they can no longer be distinguished. For this reason, it is not necessary, in practice, to take account of more than about 30 terms; and even this number is only necessary in the case of certain nebulae.

It will be seen that, by our rule, we obtain various series of terms. The first series is obtained by subtracting from Rydberg's constant successively a quarter, a ninth, a sixteenth . . of itself, so that the wave-numbers of its lines are respectively $\frac{3}{4}, \frac{8}{9}, \frac{15}{16}$ . . of Rydberg's constant. These wave-numbers correspond to lines in the ultra-violet, which can be photographed but not seen; this series of lines is called, after its discoverer, the Lyman series. Then there is a series of lines obtained by subtracting from a quarter of Rydberg's constant successively a ninth, a sixteenth, a twenty-fifth . . . of Rydberg's constant, so that the wave-numbers of this series are $\frac{5}{36}, \frac{3}{16}, \frac{21}{100}$ . . . of Rydberg's constant. This series of lines is in the visible part of the spectrum; the formula for this series was discovered as long ago as 1885 by Balmer. Then there is a series obtained by taking a ninth of Rydberg's constant, and subtracting successively a sixteenth, twenty-fifth, etc. of Rydberg's constant. This series is not visible, because its wave-numbers are so small that it is in the infra-red, but it was discovered by Paschen, after whom it is called. Thus so far as the conditions of observation admit, we may lay down this simple rule: the lines of the hydrogen spectrum are obtained from Rydberg's constant, by dividing it by any two square numbers, and subtracting the smaller resulting number from the larger. This gives the wave-number of some line in the hydrogen spectrum, if observation of a line with that wave-number is possible, and if there are not too many other lines in the immediate neighbourhood. (A square number is a number multiplied by itself: one times one, twice two, three times three, and so on; that is to say, the square numbers are 1, 4, 9, 16, 25, 36, etc.).

All this, so far, is purely empirical. Rydberg's constant, and the formulae for the lines of the hydrogen spectrum, were discovered merely by observation, and by hunting for some arithmetical formula which would make it possible to collect the different lines under some rule. For a long time the search failed because people employed wave-lengths

THE HYDROGEN SPECTRUM

instead of wave-numbers; the formulae are more complicated in wave-lengths, and therefore more difficult to discover empirically. Balmer, who discovered the formula for the visible lines in the hydrogen spectrum, expressed it in wave-lengths. But when expressed in this form it did not suggest Ritz's Principle of Combination, which led to the complete rule. Even after the rule was discovered, no one knew why there was such a rule, or what was the reason for the appearance of Rydberg's constant. The explanation of the rule, and the connection of Rydberg's constant with other known physical constants, was effected by Niels Bohr, whose theory will be explained in the next chapter.

CHAPTER V

POSSIBLE STATES OF THE HYDROGEN ATOM

It was obvious from the first that, when light is sent out by a body, this is due to something that goes on in the atom, but it used to be thought that, when the light is steady, whatever it is that causes the emission of light is going on all the time in all the atoms of the substance from which the light comes. The discovery that the lines of the spectrum are the differences between terms suggested to Bohr a quite different hypothesis, which proved immensely fruitful. He adopted the view that each of the terms corresponds to a stable condition of the atom, and that light is emitted when the atom passes from one stable state to another, and only then. The various lines of the spectrum are due, in this theory, to the various possible transitions between different stable states. Each of the lines is a statistical phenomenon: a certain percentage of the atoms are making the transition that

STATES OF THE HYDROGEN ATOM 59

gives rise to this line. Some of the lines in the spectrum are very much brighter than others ; these represent very common transitions, while the faint lines represent very rare ones. On a given occasion, some of the rarer possible transitions may not be occurring at all ; in that case, the lines corresponding to these transitions will be wholly absent on this occasion.

According to Bohr, what happens when a hydrogen atom gives out light is that its electron, which has hitherto been comparatively distant from the nucleus, suddenly jumps into an orbit which is much nearer to the nucleus. When this happens, the atom loses energy, but the energy is not lost to the world : it spreads through the surrounding medium in the shape of light-waves. When an atom absorbs light instead of emitting it, the converse process happens : energy is transferred from the surrounding medium to the atom, and takes the form of making the electron jump to a larger orbit. This accounts for fluorescence—that is to say, the subsequent emission, in certain cases, of light of a certain characteristic frequency when light of the same or greater frequency has been absorbed. The electron which has been moved to a larger orbit by outside forces (namely by the light which has been absorbed) tends to return to the smaller orbit when the outside forces are removed, and in doing so it gives rise to light belonging to a line of the atom's spectrum.

Let us first consider the results to which Bohr was led, and afterwards the reasoning by which he was led to them. We will assume, to begin with, that the electron in a hydrogen atom, in its steady states, goes round the nucleus in a circle, and that the different steady states only differ as regards the size of the circle. As a matter of fact, the electron moves sometimes in a circle and sometimes in an ellipse; but Sommerfeld, who showed how to calculate the elliptical orbits that may occur, also showed that, so far as the spectrum is concerned, the result is very nearly the same as if the orbit were always circular. We may therefore begin with the simplest case without any fear of being misled by it. The circles that are possible on Bohr's theory are also possible on the more general theory, but certain ellipses have to be added to them as further possibilities.

According to Newtonian dynamics, the electron ought to be capable of revolving in any circle which had the nucleus in the centre, or

STATES OF THE HYDROGEN ATOM 61

in any ellipse which had the nucleus in a focus; the question what orbit it would choose would depend only upon the velocity and direction of its motion at a given moment. Moreover, if outside influences increased or diminished its energy, it ought to pass by continuous graduations to a larger or smaller orbit, in which it would go on moving after the outside influences were withdrawn. According to the theory of electrodynamics, on the other hand, an atom left to itself ought gradually to radiate its energy into the surrounding aether, with the result that the electron would approach continually nearer and nearer to the nucleus. Bohr's theory differs from the traditional views on all these points. He holds that, among all the circles that ought to be possible on Newtonian principles, only a certain infinitesimal selection are really possible. There is a smallest possible circle, which has a radius of about half a hundredth millionth of a centimetre. This is the commonest circle for the electron to choose. If it does not move in this circle, it cannot move in a circle slightly larger, but must hop at once to a circle with a radius four times as large. If it wants to leave this circle for a larger one, it must hop to one with a radius nine times as large as the original radius. In fact, the only circles that are possible, in addition to the smallest circle, are those that have radii 4, 9, 16, 25, 36 . . times as large. (This is the series of square numbers, the same series that came in finding a formula for the hydrogen spectrum.) When we come to consider elliptical orbits, we shall find that there is a similar selection of possible ellipses from among all those that ought to be possible on Newtonian principles.

The atom has least energy when the orbit is smallest ; therefore the electron cannot jump from a smaller to a larger orbit except under the influence of outside forces. It may be attracted out of its course by some passing positively electrified atom, or repelled out of its course by a passing electron, or waved out of its course by light-waves. Such occurrences as these, according to the theory, may make it jump from one of the smaller possible circles to one of the larger ones. But when it is moving in a larger circle it is not in such a stable state as when it is in a smaller one, and it can jump back to a smaller circle without outside influences. When it does this, it will emit light, which will be one or other of the lines of the hydrogen spectrum according to

STATES OF THE HYDROGEN ATOM 63

the particular jump that is made. When it jumps from the circle of radius 4 to the smallest circle, it emits the line whose wave-number is $\frac{3}{4}$ of Rydberg's constant. The jump from radius 9 to the smallest circle gives the line which is $\frac{8}{9}$ of Rydberg's constant; the jump from radius 9 to radius 4 gives the line which is $\frac{5}{36}$ (i.e. $\frac{1}{4}-\frac{1}{9}$) of Rydberg's constant, and so on. The reasons why this occurs will be explained in the next chapter.

When an electron jumps from one orbit to another, this is supposed to happen instantaneously, not merely in a very short time. It is supposed that for a time it is moving in one orbit, and then instantaneously it is moving in the other, without having passed over the intermediate space. An electron is like a man who, when he is insulted, listens at first apparently unmoved, and then suddenly hits out. The process by which an electron passes from one orbit to another is at present quite unintelligible, and to all appearance contrary to everything that has hitherto been believed about the nature of physical occurrences.

This discontinuity in the motion of an electron is an instance of a more general fact which has been discovered by the extraordinary minuteness of which physical measurements have become capable. It used always to be supposed that the energy in a body could be diminished or increased continuously, but it now appears that it can only be increased or diminished by jumps of a finite amount. This strange discontinuity would be impossible if the changes in the atom were continuous ; it is possible because the atom changes from one state to another by revolution, not by evolution. Evolution in biology and relativity in physics seemed to have established the continuity of natural processes more firmly than ever before ; Newton's action at a distance, which was always considered something of a scandal, was explained away by Einstein's theory of gravitation. But just when the triumph of continuity seemed complete, and when Bergson's philosophy had enshrined it in popular thought, this inconvenient discovery about energy came and upset everything. How far it may carry us no one can yet tell. Perhaps we were not speaking correctly a moment ago when we said that an electron passes from one orbit to another " without passing over the intermediate space " ; perhaps there is no intermediate space. Perhaps it is merely habit and prejudice that makes us suppose space to be

STATES OF THE HYDROGEN ATOM 65

continuous. Poincaré—not the Prime Minister, but his cousin the mathematician, who was a great man—suggested that we should even have to give up thinking of time as continuous, and that we should have to think of a minute, for instance, as a finite number of jerks with nothing between them. This is an uncomfortable idea, and perhaps things are not so bad as that. Such speculations are for the future; as yet we have not the materials for testing them. But the discontinuity in the changes of the atom is much more than a bold speculation; it is a theory borne out by an immense mass of empirical facts.

The relation of the new mechanics to the old is very peculiar. The orbits of electrons, on the new theory, are *among* those that are possible on the traditional view, but are only an infinitesimal selection from among these. According to the Newtonian theory, an electron ought to be able to move round the nucleus in a circle of any radius, provided it moved with a suitable velocity; but according to the new theory, the only circles in which it can move are those we have already described: a certain minimum circle, and others with radii 4, 9, 16, 25, 36 . . times as large as the radius of the minimum circle. In view of this breach with the old ideas, it is odd that the orbits of electrons, down to the smallest particulars, are such as to be possible on Newtonian principles. Even the minute corrections introduced by Einstein have been utilized by Sommerfeld to explain some of the more delicate characteristics of the hydrogen spectrum. It must be understood that as regards our present question, Einstein and the theory of relativity are the crown of the old dynamics, not the beginning of the new. Einstein's work has immense philosophical and theoretical importance, but the changes which it introduces in actual physics are very small indeed until we come to deal with velocities not much less than that of light. The new dynamics of the atom, on the contrary, not merely alters our theories, but alters our view as to what actually occurs, by leading to the conclusion that change is often discontinuous, and that most of the motions which should be possible are in fact impossible. This leaves us quite unable to account for the fact that all the motions that are in fact possible are exactly in accordance with the old principles, showing that the old principles, though incomplete, must be true up to a point. Having discovered that the old principles

STATES OF THE HYDROGEN ATOM 67

are not quite true, we are completely in the dark as to why they have as much truth as they evidently have. No doubt the solution of this puzzle will be found in time, but as yet there is not the faintest hint as to how the reconciliation can be effected.

What is known about other elements than hydrogen by means of the spectroscope all goes to show that the same principles apply and that, when light is emitted, an electron jumps from an outer orbit to an inner one. But when there are many electrons revolving round a single nucleus, the mathematics becomes too difficult for our present powers, and it is impossible to establish such exact and striking coincidences of theory and observation as in the case of hydrogen. Nevertheless, what is known is sufficient to place it beyond reasonable doubt that the explanation of the spectrum of other elements is the same in principle as in the case of hydrogen. There is one case which can be tested to the full, and that is the case of positively electrified helium, which has lost one electron and has only one left. This only differs from hydrogen (as regards the movements of the electron) by the fact that the charge on the nucleus is twice as great as that on the electron, instead of being equal to it, as with hydrogen, and that the mass of the nucleus is four times that of the hydrogen nucleus. The changes which this produces in the spectrum, as compared with hydrogen, are exactly such as theory would predict.

In the present chapter, we have seen what was the conclusion to which Bohr was led as to possible states of the hydrogen atom, but we have not yet seen what was the reasoning by which he was led to this conclusion. In order to understand this reasoning, it is necessary to explain what is called the theory of quanta, of which Bohr's theory of the atom is a special case. The theory of quanta will be the subject of the next chapter.

CHAPTER VI

THE THEORY OF QUANTA

The theory that the energy of a body cannot vary continuously, but only by a certain finite amount, or exact multiples of this amount, was not originally derived from a study of the atom or the spectroscope, but from the study of the radiation of heat. The theory was first suggested by Planck in 1900, thirteen years before Bohr applied it to the atom. Planck showed that it was necessary in order to account for the laws of temperature radiation; roughly speaking, if bodies could part with their warmth continuously, and not by jumps, they ought to grow colder than they do, when they are not exposed to a source of heat. It would take us too far from our subject to go into Planck's reasoning, which is somewhat abstruse. A good account of it in English will be found in Jeans's " Report on Radiation and the Quantum-Theory," published for the Physical Society of London (1914).

Planck's principle in its original form is as follows. If a body is undergoing any kind of vibration or periodic motion of frequency ν (i.e. the body goes through its whole period ν times in a second), then there is a certain fundamental constant h such that the energy of the body owing to this periodic motion is $h\nu$ or some exact multiple of $h\nu$. That is to say, $h\nu$ is the smallest amount of energy that can exist in any periodic process whose frequency is ν, and if the energy is greater than $h\nu$ it must be exactly twice as great, or three times as great, or four times as great, or etc. The energy was at first supposed to exist in atoms or little indivisible parcels, each of amount $h\nu$. There might be several parcels together, but there could never be a fraction of a parcel. We shall see that this principle has been modified as it has been applied in new fields, so that in its present form it can no longer be stated as involving indivisible parcels of energy. But it is as well to understand its original form before considering the more recent statements of the principle.

The quantity h, which is called Planck's "quantum," is of course very very small, so small that in all the large-scale processes

THE THEORY OF QUANTA 71

observable by means of our senses there is an appearance of continuity. It is in fact so small that one unit of it is involved in one revolution of the electron in its minimum orbit round the hydrogen nucleus. It is difficult to express very large numbers in words, particularly as the word " billion " is sometimes used to mean a thousand million, and sometimes to mean a million million. If we use it to mean a million million, we may say that a billion billion times h would be a quantity just appreciable without instruments of precision. Taking the electron in its smallest orbit, h is exactly obtained by multiplying the circumference of the orbit by the velocity of the electron and multiplying the result by the mass of the electron.* In the second orbit, the result of this multiplication is $2h$, in the third, $3h$, and so on.

Planck's principle in its original form applies only to certain kinds of systems, and if rashly generalized it gives wrong results. The right way to generalize it has been discovered by Sommerfeld, but unfortunately it is very difficult to express in non-mathematical language. It turns out that the principle, in its

*Expressed in the usual C.G.S. units, $h = 6 \cdot 55 \cdot 10^{-27}$.
Its dimensions are those of action or angular momentum.

general form, cannot be stated as involving little parcels of energy; this only seemed possible because Planck was dealing with a special case. The general form requires a method of stating the principles of dynamics which is due to Hamilton. In this form, if the state of some material system is determined at any moment when we know, at that moment, how large certain quantities are (as for example the position of an aeroplane is known if we know its latitude and longitude and its height above the ground), then these quantities are called "coordinates" of the system. Corresponding to each coordinate, the system has at each moment a certain characteristic which may be called the corresponding "impulse-coordinate." In simple cases, this reduces to what is ordinarily called momentum; in a generalized sense, it may itself be called the "momentum" corresponding to the coordinate in question. It is possible to choose our coordinates in such a way that the momentum corresponding to a given coordinate at a given moment shall not involve any other coordinate. When the coordinates have been chosen in this way, the generalized quantum principle is applicable. We shall assume that such a choice has been

THE THEORY OF QUANTA 73

made. When such a choice has been made, the coordinates are said to be "separated."

The quantum-principle is only applicable to motions that are periodic, or what is called "conditionally periodic." The motion of a system is periodic if, after a certain lapse of time, its previous condition recurs, and if this goes on and on happening after equal intervals of time. The motion of a pendulum is periodic in this sense, because, when it has had time to move from left to right and from right to left, it is in the same position as before, and it goes on indefinitely repeating the same motion. Wave-motions are periodic in the same sense; so are the motions of the planets. Any motion is periodic if it can be described by means of a quantity which increases up to a maximum, then diminishes to a minimum, then increases to a maximum again, and so on, always taking the same length of time from one maximum to the next. One "period" of a periodic process is the time taken to complete the cycle from one maximum to the next, or from one minimum to the next—for example, from midnight to midnight, from New Year to New Year, from the crest of one wave to the crest of the next, or from a moment when the pendulum is at the extreme left of its beat to the next moment when it is at the extreme left.

A system is called " conditionally periodic " when its motion is compounded of a number of motions, each of which separately is periodic, but which do not have the same period. For example, the earth has a motion compounded of rotation round its axis, which takes a day, and revolution round the sun, which takes a year. There are not an exact number of days in a year, if a year is taken in the astronomical and not in the legal sense ; that is why we need a complicated system of leap-years to prevent errors from piling up. Thus when we take account of both rotation and revolution the motion of the earth is "conditionally periodic." We shall find later that the motions of electrons in their orbits, when we take account of niceties, are, strictly speaking conditionally periodic and not simply periodic. The quantum-theory in its general form applies to motions that are conditionally periodic in terms of "separated" coordinates.

We can now state the generalized quantum-principle. Take some one coordinate of the system, and imagine the motion of the system throughout one period of this coordinate divided into a great number of little bits. In

THE THEORY OF QUANTA 75

each little bit, take the generalized momentum corresponding to the coordinate in question, and multiply it by the amount of change in the coordinate during that little bit. Add up all these for all the little bits that make up one complete period. Then, in the limit, when the bits are made very small and very numerous, the result of the addition for one complete period will be exactly h or $2h$ or $3h$ or some other exact multiple of h.* No one knows in the least why this should be the case; all we can say is that it is so, in all the cases that can be tested.

In later developments we shall have occasion to consider the principle in its general form. For the present, we are only concerned with its application to the electron revolving in a circle round the hydrogen nucleus. In this case, the generalized momentum is the same thing that is called "angular momentum" in elementary dynamics; in the case of circular motion, which is the case that concerns us, it is got by multiplying the mass by the radius and the velocity. As these are all constant, there is no difficulty about obtaining the sum of little bits for a complete cycle; each little

*For the mathematical statement of the principle, see Sommerfeld's *Atombau und Spektrallinien*, 3rd ed., Chap. IV. and Appendix 7.

bit consists of the angular momentum multiplied by a little angle, and the sum of all the little bits consists of the angular momentum multiplied by four right angles; that is to say, it is obtained by multiplying the mass of the electron by the circumference (instead of the radius) of its orbit and by the velocity. By the generalized quantum-principle, this has to be h or $2\,h$ or $3\,h$ or etc. In the minimum orbit it is h; that is why no smaller orbit is possible. In the next orbit, which is four times as large, it is $2\,h$: in the third orbit, which is nine times as large, it is $3\,h$; and so on. In virtue of the quantum-principle, these are the only orbits that are possible.

We can now understand how Bohr's theory explains the lines of the hydrogen spectrum. When the electron jumps from a larger to a smaller orbit, it loses energy. A little very elementary mathematics* shows that the kinetic energy in the second orbit is a quarter of that in the first; in the third it is a ninth; in the fourth, a sixteenth; and so on. It is also very easy to show that (apart from a constant portion which may be ignored) the total energy in any orbit (potential and kinetic together) is numerically equal to the kinetic

*See Appendix.

THE THEORY OF QUANTA 77

energy, but with the opposite sign. Therefore the loss of total energy in passing from a larger to a smaller orbit is equal to the gain of kinetic energy. It follows that; if we call ε the kinetic energy in the smallest orbit, the loss of energy in passing from the second orbit to the smallest is $\frac{3}{4} \varepsilon$, the loss in passing from the third orbit to the first is $\frac{8}{9} \varepsilon$, the loss in passing from the third to the second is $\frac{5}{36} \varepsilon$, i.e. $(\frac{1}{4}-\frac{1}{9}) \varepsilon$; and so on. It will be noticed that the numbers that come here are the same as those that occurred in connection with Rydberg's constant in the preceding chapter.

The energy which is lost by the atom in one of these jumps is turned into a light-wave. What sort of light-wave it is to become is determined by the theory of quanta. A light-wave is a periodic process, and if its frequency is v, its period is a vth of a second. The generalized quantum-principle shows that, if the period of a wave is t, the energy of the wave multiplied by t must be h or an exact multiple of h; in fact, so far as observation goes, it appears to be always h. Since t is a vth of a second (when v is the frequency), it follows that the energy of the wave is hv. Also, by the principle of the conservation of energy, the energy of the wave is equal to the energy that the atom has lost.

This shows that, if ε is the kinetic energy of the electron in the smallest orbit, the wave caused by a transition from the second orbit to the first will have a frequency ν given by the equation

$$\frac{3}{4}\varepsilon = h\nu$$

For a transition from the third orbit to the first,

$$\frac{8}{9}\varepsilon = h\nu$$

For a transition from the third orbit to the second,

$$\frac{5}{36}\varepsilon = h\nu$$

and so on. Comparing these results with the empirical results set forth in Chapter IV, we see that they will agree if Rydberg's constant is equal to ε divided by h and the velocity of light. (We have to divide by the velocity of light, because in this chapter we have been speaking of frequencies, while in Chapter IV we were speaking of wave-numbers.) Now ε is easily calculated, since we know the charge on a hydrogen nucleus and on an electron, the mass of an electron, and the radius of the

THE THEORY OF QUANTA 79

minimum orbit ; also h and the velocity of light are known. It is found that the calculated value of Rydberg's constant, from these data, agrees closely with the observed value ; this was, from the first, a powerful argument in favour of Bohr's theory.

For different kinds of light, the frequency ν is different ; in the visible parts of the spectrum, it determines the colour, being smallest for red and greatest for violet. By measuring the frequencies of the different lines in the hydrogen spectrum, and multiplying each by h, we find out how much energy the atom loses in the different transitions from orbit to orbit that are possible. The terms in the spectrum are proportional to the energies in the different possible orbits, and the frequencies of the lines are proportional to the loss of energy in passing from one orbit to another. We can calculate what the different possible orbits should be from the fact that their energies must differ by an amount νh, where ν is the frequency of some line in the hydrogen spectrum. We can also calculate the possible orbits from the fact that the mass of the electron multiplied by the circumference of an orbit multiplied by the velocity in that orbit must be an exact multiple of h. These two methods lead to the same result, and thus confirm our theory.

There are, however, certain *minutiae* of the hydrogen spectrum which cannot be explained by Bohr's theory in its original form. All these, down to the smallest particular, are explained by the generalized form of the theory which is due in the main to Sommerfeld. We shall explain this generalized theory in the next chapter.

The quantity h, Planck's quantum, has been found to be involved in all the very minute phenomena that can be adequately studied. It is one of the fundamental constants to which science is led; for the present, it represents a limit of explanations, since no one knows why there is such a constant or why it is just the size it is. The limits of our explanations in any given stage of science are, while that stage lasts, brute facts; and so Planck's quantum, for the present, is a brute fact. It is involved in all very small periodic processes; but why this should be the case we do not know.

CHAPTER VII

REFINEMENTS OF HYDROGEN SPECTRUM

In Bohr's theory, the electron always moves round the hydrogen nucleus in a circle. But according to Newtonian principles, the electron ought also to be able to move in an ellipse, and the generalized quantum-principle can be applied to elliptic orbits as well as to those that are circular. It is natural to inquire whether it is possible to work out a theory that allows for elliptic orbits, and, if so, whether it will fit the facts better or worse than Bohr's original theory. It is found that, as regards the broad facts, it makes no difference whether we admit or reject elliptic orbits; in either case, the facts will accord with observation to a first approximation. There are, however, three delicate phenomena which are observed to occur, which cannot be accounted for if all the possible orbits are circles, but are to be expected if ellipses also occur. These are the following : First, there is what is called the Zeeman effect, which is an alteration produced by a strong magnetic field. Secondly, there is the Stark effect, which is produced by a strong electric field. Thirdly, there is what is called the "fine structure," which is the fact that each single line of the spectrum, when very closely examined, is found to consist of a number of almost identical lines. The explanation of the Zeeman effect is still in part incomplete, but the explanation of the other two by means of Sommerfeld's methods is as perfect as could be desired. We shall not attempt to set forth the explanation, which would be impossible without a good deal of mathematics. We shall only attempt to describe the orbits which Sommerfeld admits as possible, in addition to Bohr's circles.

If there is any reader who does not know what an ellipse looks like, he can construct one for himself by the following simple device. Tie a piece of

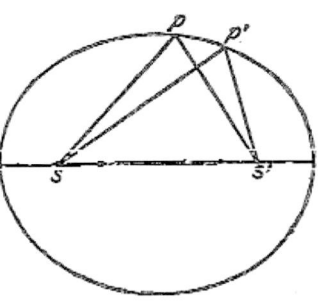

string to two pins, and stick them into a piece of paper at two points S, S' near

REFINEMENTS OF HYDROGEN SPECTRUM

enough together for the string to remain loose. Then take a pencil, and with its point draw the string taut. Any place P that the pencil will reach is on a certain ellipse, and by moving the pencil round, the whole ellipse can be drawn. The points S, S' are called its " foci." An ellipse may be defined as a curve such that, if P and P' are any two points on it, the sum of the distances of P from S and S' (the foci) is equal to the sum of the distances of P' from S and S'. In our construction, both are equal to the length of the string that we tied to the two pins. The ratio of the distance between the pins to the length of the string is called the " eccentricity " of the ellipse. It is obvious that if we were to stick the two pins into the same place we should get a circle, so that a circle is a particular case of an ellipse, namely an ellipse which has zero eccentricity. All the planets move in ellipses which are very nearly circles, whereas the comets move in ellipses which are very far removed from circles. In each case the sun is in one of the foci, and there is nothing particular in the other focus. An ellipse which is very far from being a circle can be drawn by making the distance S S' between the two pins not very much shorter than the length of the string.

There is another way of thinking of an ellipse which is also useful: it may be thought of as a circle which has been squashed. Suppose for instance that you took a wooden hoop and stood it up and put a weight on the top of it: the hoop would get squashed into more or less the shape of an ellipse. In the figure, the hoop is drawn circular, as it is before the weight is put on; then a heavy weight is put on the highest point, and the hoop takes more or less the form of the dotted curve in the

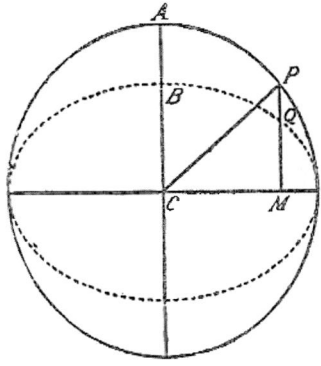

figure. The weight, which was put on at A, has made the top of the hoop sink to B. The hoop is supposed to be fastened, like a wheel, on to an axle in the middle, C. An ellipse can be obtained from a circle which is standing upright by diminishing all vertical distances in a certain fixed proportion; that is to say, if P is any point of the circle, which is at a height P M above the level of the axle, we go down to a point Q below P, such that the height Q M bears a fixed ratio to P M, the same, of course, as the ratio of B C to A C. The ratio of P Q to Q M is also of course the same for any point of the curve, and equal to the ratio of A B to B C. We will call this the amount of " flattening " of the ellipse. This is not a recognized expression, but will prove convenient for our purposes. To state the whole thing precisely : Given a circle, imagine it to be stood upright, like a wheel, with an axle through the centre. Then lower each point in the top half of the wheel by a fixed proportion of its height above the level of the axle, and raise each point in the bottom half in the same proportion ; the proportion of the lowering to the final height (or of the raising to the final depth, in the lower half) we will call the amount of " flattening " in the ellipse. That is to say, if A B is half of B C (and P Q half of Q M), the amount of flattening is a half ; if A B is a third of B C, the amount of flattening is a third ; and so on.

We can now explain what are the ellipses which are possible for the electron in a hydrogen atom.

In the figure, E represents the electron, S represents the nucleus, which is in a focus of the ellipse. N is the point where the electron is nearest to the nucleus. F the point where it is farthest from it, C the centre of the ellipse, which is half way between N and F. There are now two periodic characteristics of the orbit, instead of only one, as in the case of the circle. The first periodic characteristic is, as before, the angle which S E makes with S F. The other is S E, the distance of the electron from the nucleus. This grows continually smaller while the electron is travelling from

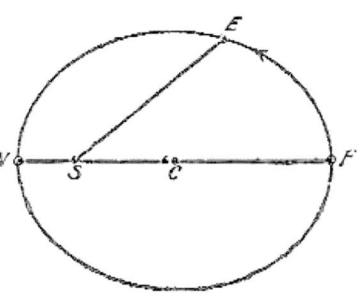

F to N, and then continually larger while it is travelling from N to F. As there are two periodic characteristics of the orbit, the general quantum-theory will give two conditions that the orbit must fulfil, instead of only one. It is impossible to explain the process by which the results are obtained, but the results themselves are fairly simple.

The first quantum condition is very much the same as in the case of circular orbits. Take the mass of the electron, its velocity at N (where it is nearest to the nucleus S), and the circumference of the circle whose radius is S N, and multiply these three together; the result must be an exact multiple of h, say $n\,h$. The second quantum condition determines how much the ellipse departs from a circle; it states that there is a second whole number n' (the second quantum number), such that $\dfrac{n'}{n}$ is the amount of flattening in the sense defined a moment ago. The second number n' may be zero; we then obtain Bohr's case of circular orbits. If it is not zero, the electron moves in a more or less eccentric orbit.

It turns out that, apart from niceties, the energy of an electron in its orbit, and therefore the spectral lines corresponding to jumps from one orbit to another, do not depend upon the separate numbers n and n', but only upon their sum $n+n'$. The result is that the lines to be expected, apart from niceties, are the same as on Bohr's original theory of circular orbits. If the matter were to end here, we might seem to have had a lot of trouble for nothing. Even then, however, we could have drawn a useful lesson from the theory of elliptic orbits. There are, as we shall see, certain facts which are explained by elliptic orbits and not by circular orbits, but these facts are mostly recent discoveries, and might easily have remained unknown for some time longer. In that case, Bohr's original theory would have accounted admirably for all the known facts, and there would have seemed to be very strong grounds for accepting it. Yet the theory of elliptic orbits would have accounted for the facts just as well, so that there would have been no way of deciding between them. This illustrates what is sometimes forgotten, that a theory which explains all the known relevant facts down to the minutest particular may nevertheless be wrong. There may be other theories, which no one has yet thought of, which account equally well for all that is known. We cannot accept a theory with any confidence merely because it explains what is known. If we are to feel any security, we must be able to show that no other theory would account for the facts. Sometimes this is possible, but very often it is not. Poincaré advanced a proof that the facts of temperature radiation cannot be explained if we assume that radiation is a continuous process, and that any possible explanation must involve sudden jumps such as we have in the quantum theory. His argument is difficult, and it is possible that it may not ultimately prove wholly cogent. But it affords an instance of that further step without which scientific hypotheses must remain hypothetical. In our case, fortunately, there is evidence that elliptic orbits actually do occur when an electron moves round a hydrogen nucleus. That is to say, there is evidence that this hypothesis explains certain facts which the hypothesis of circular orbits cannot explain. But although the agreement between theory and observation is astonishingly close, it cannot be said that we have yet reached the stage where we can be quite certain that no other theory would account for the facts.

All the broad facts in the spectrum depend upon the sum of the two quantum-numbers, $n + n'$, not on either separately. We therefore classify orbits by this sum. We thus arrive at the following possible orbits.

1st case. $n + n' = 1$. Since n cannot be zero (because if it were the electron would fall into the nucleus), this gives only one possibility, namely $n = 1$, $n' = 0$. When

$n' = 0$, there is no flattening, and the orbit is a circle. Thus this first case is that of Bohr's minimum circle.

2nd case. $n + n' = 2$. Here there are two possibilities, namely $n = 2$, $n' = 0$ and $n = 1$, $n' = 1$. The first of these gives Bohr's second circle; the second gives an ellipse in which there is a unit amount of flattening, that is to say, the ellipse is half as high as it is broad.

3rd case. $n + n' = 3$. Here there are three possibilities, namely: (a) $n = 3$, $n' = 0$; this gives Bohr's third circle. (b) $n = 2$, $n' = 1$; this gives an ellipse in which the amount of flattening is a half, that is to say, the ellipse is two-thirds as high as it is broad. (c) $n = 1$, $n' = 2$; this gives an ellipse in which the amount of flattening is two, that is to say, the ellipse is a third as high as it is broad.

In the fourth case, $n + n' = 4$, there are our possibilities; and so on. The breadth of the ellipse depends only upon $n + n'$, so that all the possible ellipses under one head have the same breadth. The energy also, apart from niceties, depends only upon $n + n'$*.

Of the three sets of facts which show that elliptical orbits must occur, we shall pass by

*For a full mathematical treatment of the above topic, see Sommerfeld, *op. cit.*, pp. 286–297

the Zeeman effect (which shows how magnetism splits one line into three, or sometimes more) and the Stark effect (which shows the influence of an electric field). The third, however, is so interesting that it cannot be omitted, since it shows that the electron, in so far as it obeys ordinary dynamical laws, follows the principles of Einstein in preference to those of Newton.

Very careful observation shows that the lines in the spectrum which we have hitherto treated as single really consist of two (and in other cases three or more) separate lines very close together. This suggests that two different orbits which give the same value of $n + n'$ do not produce *exactly* the same line in the spectrum when an electron jumps to or from them. The phenomenon is more noticeable in the case of other elements than in that of hydrogen, for reasons which the theory explains. Fortunately on this point our theory is able to tell us a good deal about other atoms; but in what follows we shall confine ourselves to hydrogen.

The mathematical argument which shows that the energy of the electron in its orbit only depends upon $n + n'$ proceeds on Newtonian principles; more particularly, it treats the mass of the electron as constant. But in the modern theory of relativity, the mass of a body is increased by rapid motion. This increase is not noticeable for ordinary velocities, but becomes very great as we approach the velocity of light, which is a limit that no material body can quite reach. Readers may remember that Einstein's theory of gravitation has been confirmed by two facts which remained inexplicable on Newtonian principles. One is the fact that light bends by a certain amount (double what Newtonian principles allow) when it passes near the sun, which has been verified in two eclipses. The other is the fact which is called the motion of the perihelion of Mercury, which had long been known to astronomers without their being able to find any way of accounting for it. It is the analogue of this fact that concerns us. Mercury, like the other planets, moves in an ellipse with the sun in a focus ; it is sometimes nearer to the sun and sometimes further from it. Its " perihelion " is the point of its orbit which is nearest to the sun. Now it has been found by observation that, when Mercury has gone once round the sun from its previous perihelion, it has not quite reached its next perihelion ; that is to say, it has to

REFINEMENTS OF HYDROGEN SPECTRUM 93

go a little more than once round the sun in passing from one occasion when it is nearest the sun to the next. This of course shows that its orbit is not quite accurately an ellipse. There is supposed to be a similar phenomenon in the motions of the other planets, but it is too small to be observed ; in the case of Mercury it is just large enough to be noticeable. Einstein's theory of gravitation, but not Newton's, explains why it exists, and why it is just as large as it is ; it also explains why the effect in the case of the other planets is too small to be observed. In order to be noticeable, the orbit must depart fairly widely from a circle, but the orbits of all the planets except Mercury are very nearly circular.

In the case of the electron in the hydrogen atom, as we have seen, the possible orbits which are not circles are very markedly elliptical. This makes the effect which has been noticed in the case of Mercury very much more pronounced in the case of the electron. Moreover, since the velocity of the electron in its orbit is much greater than that of the planets, there is a much more noticeable effect of the increased velocity, when the electron is near the nucleus, in increasing the mass. This causes a quite appreciable effect of the same sort as the motion of the perihelion of Mercury. That is to say, the electron makes a little more than one complete revolution between one occasion when it is nearest to the nucleus and the next occasion when this happens. It is found that this accounts for the fine structure of the spectral lines, though it would be impossible to set forth the explanation in non-mathematical language. It is curious that, although the quantum theory is something quite outside traditional dynamics, everything unaffected by this theory proceeds exactly according to the very best principles of non-quantum dynamics, that is to say, according to Einstein rather than Newton. The proof is in this fact of the fine structure.*

*The mathematical theory of the fine structure will be found in Sommerfeld, *op. cit.*. Chap. VIII. The explanation of the motion of the perihelion in the above is not, properly speaking, the same as in the case of Mercury'; the latter depends upon the general theory or relativity and Einstein's new law of gravitation, while the former depends only upon the special theory of relativity.

CHAPTER VIII

RINGS OF ELECTRONS

When we come to atoms that have more than one electron, we can no longer work out the mathematics in the same complete way as we can in the case of hydrogen and positively electrified helium. We shall see in the next chapter, however, that X-ray spectra (which are a very modern discovery) tell us a great deal about the inner rings of electrons in complex atoms, while optical spectra continue to tell us a good deal about the outer ring. As we travel up the periodic table, the first element in each period, which is an alkali, has only one electron in the outermost ring; accordingly we might expect this one electron to move more or less as the hydrogen electron does, since the positive charge on the nucleus exceeds the negative charges on the inner electrons by just the amount of the charge on an electron or a hydrogen nucleus, and the inner electrons may be expected to be never very near the outer electron, as distances go within an atom. This would lead us to look out for a spectrum, in the case of an alkali, more or less similar to that of hydrogen ; and in fact this is found to be the case. Some inferences can be drawn from the fact that in all series spectra Rydberg's constant makes its appearance. There can be no doubt that the quantum theory applies, and that the orbit of an electron (as in the case of elliptical orbits in hydrogen) is in general determined by two quantum numbers, both of them whole numbers which are usually small.

There is, however, considerable uncertainty about the arrangement of the electrons when there are more than one.

Already with helium, which has only two electrons, complications arise. There are two complete systems in the helium spectrum, each such as one might expect to constitute the whole spectrum of an element. This leads Bohr to the conclusion that there are two possibilities for the stable state of the second electron, in one of which it moves in an orbit similar to that of the first, while in the other it moves in an orbit considerably larger than that of the first. These two states would not be related as are the different possible orbits

RINGS OF ELECTRONS

in the hydrogen atom; that is to say, an electron left to itself would never jump from the larger to the smaller orbit. They are both final states, after all the jumps have been made. The atom cannot pass from one to the other directly, but only by a roundabout process. When both electrons move in similar minimum orbits, they cannot be in the same plane. Originally it was assumed, merely in order to try simple hypotheses first, that the electrons in an atom all moved in the same plane. This hypothesis has had to be abandoned, and it is now believed that even the electrons constituting one ring are in different planes. In fact it is suggested that, in an inert gas, the eight electrons constituting the outer ring are arranged more or less like the eight corners of a cube. But according to Bohr even this hypothesis is still too simple.

It will be remembered that, when we were dealing with elliptic orbits in the hydrogen atom, we found that the two quantum numbers n and n' were not individually so important as their sum, $n + n'$. We call this the "total quantum number." Although we cannot calculate in detail the paths of electrons in other atoms, we can see that there will still be a "total quantum number," the sum of two partial quantum numbers, which will determine the most important features of the orbit. Rings of electrons will be sets having the same total quantum number. If their two partial quantum numbers severally are the same, their orbits will have the same shape ; if not, the orbits of some will be much more eccentric than those of others. It may happen that the orbit of an electron belonging to an outer ring is so eccentric that at moments it penetrates within an inner ring, just as a comet which is usually very distant from the sun may for a short time be nearer than any of the planets. When an electron penetrates in this way into regions thoroughly settled by other electrons, all of which are repelled by it, the effect must be very disturbing. Comets produce no great disturbance in the solar system, because their mass is very small ; but electrons are all equal, not only as to their mass, which is less important, but as to their electric charge, which is what governs their motions. It seems as if an atom must be somewhat uncomfortable, and have anything but a harmonious family life, if it is subject to such irruptions several billions of times in every second. However, apparently it gets used to them, and learns to adjust itself.

RINGS OF ELECTRONS

The phenomena of the optical spectrum are produced by disturbances in the outer ring of electrons, i.e. when one of the outer electrons has been moving in an orbit which is larger than the normal orbit of an electron in the outer ring, and suddenly jumps to this normal orbit or to some intermediate one. But X-rays arise from disturbances in the inner rings of electrons. If an electron is torn away from the inner regions of an atom, it will soon be replaced by some electron which was formerly in the outer ring; there is a vacant place near the nucleus, and any electron that can will seize the chance to occupy it. The amount of energy radiated out in waves when this occurs is very great, and therefore the frequency of the waves is very great. X-rays only differ from ordinary light-waves by their great frequency, so that the emission of X-rays is just what might be expected under such circumstances. This is why X-rays give us so much information about the inner rings.

Bohr[*] has given a table setting out his theory of the way the electrons are arranged in the various inert gases, each of which has its outer ring as full as it will hold until there

[*] *Op. cit.* p. 113.

are other electrons outside it. The helium atom, in its commoner form, he supposes to contain two electrons moving in circles, each with the same total quantum number, namely 1, as the minimum circle in hydrogen. There is, however, as we saw, another form of helium, in which one of the electrons moves in an eccentric orbit. In the next inert gas, neon, there are 10 electrons, two in the inner ring and eight in the outer. The two in the inner ring, according to his table, remain as in helium, but of the outer eight four are moving in circles and four in ellipses. This and the other figures in his table apply, of course, to the atom in its most compressed state, the state to which it tends when it is let alone, the state corresponding to the minimum circle in hydrogen. Argon, which comes next with 18 electrons, has its two inner rings as in neon, but has eight electrons in a third ring. Partly from spectroscopic considerations, partly on grounds of stability, Bohr maintains that these eight outer electrons none of them move in circles, but are divided into two groups of four, the first group moving in orbits of very great eccentricity, the second in less eccentric orbits. The first group of four will, at moments, penetrate inside the first ring. It is assumed

RINGS OF ELECTRONS

that the two inner rings are definitely completed as soon as we reach neon, but that the later rings are not completed so quickly. For reasons which we explained in Chapter III, the periods containing a great many elements in the periodic table are best explained by assuming that the change from one element to the next is not always in the outermost ring, but is sometimes in the next ring, or even (in the case of the rare earths) in the next but one. According to these principles, krypton, which is the 36th element, and so has 18 more electrons than argon, is not to have the whole 18 in its outer ring. Only 8 are to be in the outer ring; the remaining 10 are added to the third ring, which is to have eighteen electrons, six in orbits like one previous group of four, six in orbits like the other previous group of four, and six in circles. The eight outer electrons are again divided into two groups of four, one group exceedingly eccentric (more so than any in argon), and the other group somewhat less so. Passing to Xenon, the 54th element in the periodic table, the first three rings are as in krypton, the fourth ring has 18 electrons instead of 8, six in each of the groups that previously had four, and six in orbits that are not circles, but have only a small eccentricity. As we saw in connection with hydrogen in the previous chapter, as the total quantum number increases, the number of possible orbits increases. When the total quantum number is one, there is only one possibility (a circle); when 2, there are two; when 3, there are three, and so on. This does not mean that there can be only one orbit whose total quantum number is one; it only means that any orbit whose total quantum number is one must be a circle of a certain size. There may be (except in hydrogen there are) two electrons moving in circles of this size, but in different planes. Similarly in the other cases. As we travel up the series of total quantum numbers, more and more eccentric orbits become possible; circles always remain possible, but the number of possible types of ellipses increases by one at each step. When the total quantum number is three (third ring), the ratio of the breadth to the height may be 3 or 2 or 1. (The ratio 1 corresponds to a circle). When it is four (fourth ring), the ratio may be 4 or 3 or 2 or 1; and so on. When the breadth is very much greater than the height, the orbit is very eccentric. Bohr holds that in each ring the more eccentric orbits are filled first, and the less eccentric later ; he bases this view on considerations of stability, because we always have to account for the fact that the system of electrons does not break down more often than it does.

In accordance with this principle, the outer (fifth) ring in Xenon is to have eight electrons divided into two groups of four, the first group having the most eccentric orbits possible at this stage (length five times breadth), the second group having the next most eccentric orbits (length five times half the breadth). For convenience, we are speaking as if the orbits of the electrons were still ellipses and circles, but of course this is only very roughly true when we have to deal with a crowd of electrons which all have to dodge each other. It is only true to the same degree that a person walking along Oxford Street in the afternoon walks in a straight line ; a straight line gives the general direction of his movement, but he is always deviating from it to get out of people's way. Similarly the electrons, when they come close together, repel each other violently, and shove each other out of the smooth circular or elliptical course. But for general descriptive purposes it is convenient to ignore this. What we can hope to find out about the electrons is the quantum-numbers of their orbits, because these determine the spectral lines. But we cannot hope, with our present mathematical knowledge, to calculate exactly the orbit of an electron with two given quantum numbers, although we can see in a general way what sort of orbit it must be. This is to be borne in mind when, for brevity, we speak of ellipses and circles in connection with atoms that have a great many electrons.

Between Xenon and Niton comes the period of 32 elements, so that in constructing a model of the Niton atom in its normal state we have to find room for 32 new electrons. This is done as follows : the first three rings remain unchanged ; the fourth is augmented to contain 32 electrons, 8 in each group that previously held 6, and 8 in circular orbits ; the fifth ring is increased from 8 electrons to 18, of which there are 6 in each group that previously held 4, and 6 in a new group of slightly eccentric orbits ; the sixth ring contains 8 electrons, four moving in very eccentric orbits (length six times breadth), the other four in less eccentric orbits (length three times breadth). It would of course be possible to go on constructing models of atoms with larger numbers of electrons, but after niton only five more elements are known, and they are breaking down through radio-activity. It seems therefore that the series stops where it does because heavier atoms would not be stable. However, since new elements are discovered from time to time, we cannot be sure that no element heavier than uranium will ever be discovered. It would therefore be rash to set to work to prove that such elements are impossible.

It must not be supposed that the above models of complicated atoms have the same degree of certainty as the theory of the hydrogen atom. They are as yet in part speculative. But it is in the highest degree probable that the models give a more or less correct general picture of the way the electrons behave when the atoms in question are in their most stable state. The emission of light and X-rays occurs when one electron makes a transition towards the most stable configuration, which is the one intended to be described by the models we have been considering. Absorption, on the contrary, takes place when there is a transition away from the most stable configuration under the influence of outside forces.

CHAPTER IX

X-RAYS

Everybody knows something about X-rays, because of their use in medicine. Everybody knows that they can take a photograph of the skeleton of a living person, and show the exact position of a bullet lodged in the brain. But not everybody knows why this is so. The reason is that the capacity of ordinary matter for stopping them varies approximately as the fourth power of the atomic number of the elements concerned. Thus carbon, whose atomic number is 6, is 1296 times as effective as hydrogen in stopping X-rays; oxygen, whose atomic number is 8, is 4096 times as effective as hydrogen; nitrogen, whose atomic number is 7, is 2401 times as effective as hydrogen; calcium, whose atomic number is 20, is 160,000 as effective as hydrogen. The human body consists mainly of carbon, oxygen, nitrogen and hydrogen, but the bones consist mainly of calcium. Consequently

X-rays which go through the rest of the body easily are stopped by the bones, with the result that we get a photograph of the skeleton. Lead, of which the atomic number is 82, is about 45 million times as effective as hydrogen, and about 280 times as effective as calcium ; so it is no wonder that bullets come out clearly in X-ray photographs.

In this chapter, we shall be concerned with the physical nature of X-rays, not with their application to medicine.

When swiftly moving electrons strike ordinary matter, which happens in the case of so-called " cathode-rays " and " β-rays," they give rise to X-rays, which were discovered by Roentgen in 1895. It was not known until 10 years later whether these rays were longitudinal or transverse ; then Barkla showed that they are transverse, like light, and it is now known that they only differ from light by their very much greater frequency. When a body is hit by X-rays, it gives out X-rays itself, which are called " secondary X-rays." These in turn give rise to " tertiary X-rays." The X-rays emitted by a body are of two sorts, partly mixed and having no particular relation to the body which emits them, partly characteristic of the body. It is only the latter that can be said to have a spectrum belonging to the substance of which the body is composed. The characteristic X-rays emitted by an element, when analysed, are found to consist of only a few sharp lines, giving a very simple spectrum, which varies in a perfectly regular manner with the atomic number. Unlike the optical spectra, the X-ray spectra of different elements are closely similar, with an increase of frequency in corresponding lines as the atomic number increases. Broadly speaking, there are three sets of lines, the K, L, and M lines as they are called, which make up the X-ray spectra; but technical difficulties make it impossible to observe more than two in one element. None can be observed in very light elements; the K-line cannot be observed in very heavy elements, and the M-lines can only be observed in very heavy elements. But this is fully accounted for by the difficulties of observation. X-ray spectra can only be observed by means of suitable crystals, and the observations are limited by the crystals that are available. There is every reason to believe that, if we could invent suitable apparatus, we should find that all three lines exist throughout the series of elements. They are, in fact, roughly the same as the principal lines in the hydrogen spectrum, which in that case fall in the optical region. The frequency of each line increases very nearly as the square of the atomic number, as we pass from one element to another. Each line corresponds to a transition from one ring of electrons to another.

What may be supposed to happen when X-rays are excited is closely similar to what happens when visible rays are excited. An electron, passing in the kathode stream (which consists of swiftly moving electrons), penetrates into the inner rings of electrons, and manages to knock out one of the electrons in an inner ring. The resulting state of affairs is unstable, and presently the outer rings supply an electron to the vacant place in the inner ring. The result is that the atom loses energy, which spreads out in a wave just like a light-wave. But when heavy atoms are concerned, the great charge on the nucleus causes the time of revolution of the nearer electrons to be much less than in the case of hydrogen. Roughly speaking, the number of revolutions per second in an orbit having given quantum numbers will increase as the square of the atomic number of the element concerned. It follows that this applies also to the difference of energy between two different rings, and therefore (by Planck's principle) to the frequency of the corresponding spectral line; for, by Planck's principle, when an electron jumps from one ring to another, the frequency of the corresponding spectral line is obtained by dividing the loss of energy by h. Thus roughly speaking we should expect X-ray spectra to give the same lines for different elements, only with frequencies that increase as the square of the atomic number; and in fact this is what we do find. It is because the frequency increases so fast as we go up the periodic table that the inner rings of the later elements give lines in the X-ray spectrum instead of the optical spectrum.

X-ray spectra do not occur, as a regular thing, in the form of absorption spectra, and in this they differ from optical spectra. It is worth while to understand why this is. When ordinary light, of a frequency which an element is capable of emitting, passes through a gas composed of the element, the element absorbs all or some of it, though light of other frequencies passes through freely. The reason is that light corresponding to a spectral line of the element supplies just the quantum required to move an electron from an inner to an outer ring. The energy of the light-wave is used up in doing this. The electrons involved in optical spectra are only those in the outer ring ; in a case of absorption, they are moved still further out into an empty region, from which they may return at some later time in a case of fluorescence. But in X-ray spectra the electrons concerned are those in the inner rings. When one of these is fetched out by a passing electron, it cannot settle in an outer ring, because the outer rings are already occupied by electrons. Each of the electrons that it passes on the way out repels it, and gives it (so to speak) an extra shove. The result is that it cannot rest in an outer ring, unless by some exceptional stroke of luck, but has to go wandering off into space. The energy involved in such a journey is not tied down to certain amounts, like the energy involved in passing from one possible orbit to another. Its place in the inside is taken by one of the outer electrons, while the outer ring remains one electron short until it has a chance to help itself from some other atom or by means of some free electron.

We saw in Chapter II that what is called the atomic number of an element is more important than the atomic weight. The atomic number represents a fundamental property of the atom, namely the positive charge on the nucleus ; an atom with such-and-such an atomic number has a charge on the nucleus which is such-and-such a number of times the charge on the hydrogen nucleus, or the opposite charge on the electron. It follows that an atom in its neutral state, i.e. when it is unelectrified, has a number of electrons round the nucleus which is the same as its atomic number. But atomic weight had the prestige of tradition as the characteristic by which atoms should be arranged in a series, and the few cases (four in all) where the periodic table inverts the order of atomic weights were felt to be annoying. X-ray spectra, however, have given a decisive victory to the classification by atomic numbers. We saw that different elements have very similar X-ray spectra, except that the frequencies of corresponding lines increase as the square of the atomic number (approximately) as we pass from element to element. This law is fulfilled just as exactly in cases in which the atomic weight would invert the order as it is in other cases. This is what theory would lead us to expect, if each step up the periodic series makes an increase of one in the positive charge on the nucleus ; and on any other hypothesis it seems scarcely possible. The X-ray spectra, therefore, afford a very powerful argument in favour of Rutherford's general conception of the way atoms are constructed, as well as in favour of the theory of quanta as the explanation of spectral lines.

The law of X-ray spectra is the same as the law of optical spectra, namely that, if ν is the frequency of a line in the spectrum (i.e. the number of waves per second), and h is Planck's quantum, h multiplied by ν is the energy lost by the atom in the transition which gives rise to the line in question. There are three principal sets of lines in X-ray spectra, called, respectively, the K, L, and M lines. For any given atom, the K lines have the greatest frequency and the M lines the least. The K lines represent a transition by an outer electron to the inmost ring, an L line represents a transition to the second ring, and an M line to the third. Each line, closely examined, is found not to be single, but to consist of several neighbouring lines, corresponding to different starting-points for the electron, but all having the same end-point. Since we can observe the frequencies of the different lines, we can infer from the X-ray spectra what are the differences between the energies of electrons in different rings. Everything confirms the theory of the structure of atoms which was suggested by the hydrogen spectrum and the facts upon which the periodic table is based.

Another very instructive fact which emerges from the study of X-ray spectra concerns the "fine structure," of which, as we saw in Chapter VII, the explanation is to be sought in the substitution of Einstein's principles for Newton's. In the case of hydrogen, the different lines of the fine structures are so near together that accurate measurements of their distance apart are very difficult. But the distance between them, as we pass to later elements, ought to increase roughly as the fourth power of the atomic number, so that measurements become much easier for high atomic numbers. On this point, the empirical evidence obtained from X-ray spectra agrees closely with the theory developed by Sommerfeld. This theory depended, it will be remembered, on the fact that, according to the doctrine of relativity, an electron which moves in an eccentric orbit has to go rather more than once round its orbit before getting back to the point at which it is nearest to the nucleus. The X-ray observations establish this theory much more firmly than is possible by the help of optical spectra alone.

It must be understood that, so far as quantum numbers are concerned, the actual orbits of electrons in atoms that have many rings are the same as the possible orbits of the one electron in the hydrogen atom. The spatial dimensions are not the same; the radius of the minimum circle, roughly speaking, varies inversely as the atomic number, so that in uranium it might be expected to be about 92 times smaller than in hydrogen. The velocity of the inside electron in its minimum orbit varies roughly as the atomic number and the number of revolutions per second roughly as the square of the atomic number. But the radius multiplied by the velocity is independent of the atomic number. To a first approximation, the mass of the electron multiplied by its velocity multiplied by the circumference of its orbit (when it moves in a circle) will always be h in the inmost ring, $2h$ in the second, $3h$ in the third, and so on. The important thing to know about an orbit is what are its quantum numbers, i.e. what multiples of h are involved. This is just as true in regard to X-ray spectra as in regard to optical spectra.

It will be seen that the electron in a hydrogen atom has, in a certain sense, more freedom than one of the many electrons in heavier atoms. There is less overcrowding, and more room for migration. Under the influence of incident light, the hydrogen electron can move out to a larger orbit; presently, when the light is gone, it can return again. But an electron in one of the inner rings of a heavy atom cannot remove at will to another orbit. If it is forced to leave its orbit, it has to leave the atom altogether. The other paths which the quantum-theory permits are occupied, until we get to a considerable distance from the nucleus, whereas in hydrogen they are vacant. Paths that have large quantum numbers, though possible in theory, cannot occur in practice, at any rate in the laboratory, because they are so large that they would cause the electron to get into the region of other atoms. In certain nebulae, where matter is almost inconceivably tenuous, the spectrum shows that electrons can travel round hydrogen nuclei in orbits whose total quantum number is as large as 30. But even in the nearest approach to a vacuum that we can create artificially there are still too many atoms for such large orbits to be possible. That is why there is a limit, in practice, to the number of lines in the spectrum of an element, although, in theory, the number of possible lines is infinite.

CHAPTER X

RADIO-ACTIVITY

Shortly after the discovery of X-rays, the world was startled by the discovery of radio-activity. The discoverer was the French physicist Becquerel. What first led him to the discovery was the fact that a very sensitive photographic plate was put away in a dark cupboard with a piece of uranium, and was found afterwards to have photographed the uranium in spite of the complete darkness. On investigating this remarkable phenomenon, Becquerel found that the rays which produced the photograph came from the uranium itself, and did not depend upon any previous exposure to light, as is the case with fluorescent substances. Uranium was found to be able to produce rays out of itself apparently indefinitely, and these rays were very powerful. At first the discovery was upsetting. It *seemed* to go against the conservation of energy, because the energy radiated by the uranium was to all appearances created out of nothing. This turned out not to be the case; the energy, as we shall see, comes out of the nucleus of the uranium atom. But something equally astonishing was found to happen: in radio-activity one element turns into another. Throughout the middle ages, chemists had tried in vain to transmute elements; the impossibility of doing so seemed to be one of the most certain results of chemistry. This has proved to be a mistake; in radio-activity atoms of one element throw out particles from the nucleus and become atoms of another element.

Radio-activity is associated in popular thought with radium, but in fact the discovery of radium was caused by that of radio-activity, not vice versa. Monsieur and Madame Curie, who were working under Becquerel, observed that pitchblende, from which uranium is obtained, is more radio-active than pure uranium. They inferred that it must contain some very radio-active constituent, much more active than uranium. The search finally led Madame Curie to the new element radium. Since then, a number of new radio-active elements have been discovered. Sommerfeld (*op. cit.* p. 56) enumerates forty of them, and there is no reason to suppose the list complete.

Before going into the process by which a radio-active atom disintegrates, let us consider the rate at which different radio-active substances decay. The atoms of a radio-active substance are like a population which has a certain death-rate; in a given time, a given percentage of them die, and are born again as atoms of a different substance. But they are not endowed, like human beings, with a certain span of life. Some live a very short time, and some a very long time; the old ones are no more liable to death than the young ones. So far as we can tell, any population of atoms of a given radio-active element will lose a certain proportion in a given time, quite regardless of the question whether the atoms are old or young. It is customary to measure the rapidity of disintegration by the length of time that it takes for half of a given collection of atoms to die. This period varies enormously from one substance to another. Uranium, which is only very slightly radio-active, takes 4500 million years, in its most stable form, for half its atoms to decay. The first product of their disintegration is a substance of which half decays in just under 24 days; this breaks down into a substance for which the period is less than a minute and a quarter; the next substance has an uncertain period, estimated at two million years; at this stage, two different products may be formed, one of which in turn becomes radium, of which the period is 1580 years, while the other becomes Protoactinium, of which the period is 12000 years, the next product being actinium. Radium gives rise to the inert gas Niton (also called Radium-emanation), for which the period is a little less than 4 days. The end of both series is a form of lead, which, so far as we know, is not radio-active at all. There is a separate family starting from Thorium (which has the atomic number 90); this also ends in a form of lead (atomic number 82). Some radio-active products decay so fast that half of them die in a tiny fraction of a second. The shortest time is estimated at a hundred-thousandth of a millionth of a second, but this is more or less conjectural.

It must not be supposed that, if half the atoms of a substance die in a certain period, all will die in double that period. After half are dead, only half as many are left to die; of these half will die in the next period. Thus to take radium: Given a certain number of atoms of radium, half decay in 1580 years, and half are left at the end of that time. In the next 1580 years, half of that half will decay, and a quarter of the original number will be left; at the end of a third period of 1580 years, an eighth of the original number will be left, and so on.

The exact circumstances which make a radio-active atom break up are not known; we only know statistical averages. We have to suppose that the nucleus is in more or less unstable equilibrium, and may be disintegrated at any moment by some chance which comes, on the average, to a certain proportion of the atoms in any given period. We are in the same position as we should be in with human populations if we could observe the death-rate, but were quite unable to observe the various diseases of which people die. One point in which radio-active substances differ from human populations is that, at the beginning of the series, we have two substances, uranium and thorium, which sometimes die but are never born, so far as our knowledge extends, while at the other end we have three kinds of lead, which are born but apparently never die. Thus the heaviest elements in the periodic series are continually breaking down, and no process is known by which they can be built up again. There may at one time have been many elements with a structure more complex than that of uranium, which have broken down so that whatever traces of them are left in the universe have not been discovered by us. Radio-activity is one of those processes of degeneration (in a certain technical sense) to which no converse process of regeneration is known. We see complex atoms breaking up, and it is natural to suppose that there are (or have been) circumstances under which they are put together out of simpler atoms. But no trace of any such circumstances has been discovered. In this respect, as in some others, the universe *seems* like a clock running down, with no mechanism for winding it up again. All the uranium in the world is breaking down, and we know of no source from which new uranium can come. Under these circumstances it seems strange that there should be any uranium. But if, like some insects, we lived only for a single spring day, we should think it strange that there should be any ice in the world, since we should find it always melting and never being formed. Perhaps the universe has long cycles of alternate winding-up and running-down ; if so, we are in the part of the cycle in which the universe (or at least our portion of it) runs down. Everything pleasant is associated with this running down, because it is only this process that liberates energy for the purposes that we regard as useful. It is time, however, to return from these speculations to the mechanism of radio-activity.

When a substance is radio-active, it emits one or more of three kinds of rays, which are called respectively α-rays, β-rays, and γ-rays. It has been found that α-rays consist of particles, each of which is the nucleus of a helium atom ; β-rays also consist of particles, but in this case they are electrons ; γ-rays do not consist of particles, but are of the nature of light-waves, only with a very much higher frequency, higher even (sometimes 20 times higher) than that of X-rays. We need not further consider the γ-rays, which do not concern the transformations of the nucleus. It is the α-rays and β-rays that produce the results in which we are interested. We will begin with the α-rays.

The α-rays are compounded of α-particles, which are nuclei of helium, and thus have a positive charge double that of the hydrogen ns, and a mass (or weight) four times that of the hydrogen nucleus. They are shot out with a velocity which may reach to nearly a tenth of the velocity of light. Since they have a double positive charge, they attract electrons, and therefore it is not surprising that they tear away electrons from any atoms they may meet, and so cause the matter on their path to become positively electrified. When they have captured the two electrons that they desire, they become ordinary unelectrified helium atoms. Being small and heavy and swift, they have great power of penetration through ordinary matter. They come from the nucleus of the atom, which thus loses two units of positive electricity, and therefore is moved down two places in the periodic series. At the same time the atomic weight diminishes by four, because the helium nucleus is four times as heavy as the hydrogen nucleus. If the α-particle left the electrons of the atom undisturbed, there would be an excess of two electrons in the atom after its departure ; but in fact it generally tears away at least two electrons as it goes. If it loosens more than two, the atom will become positively electrified until it can annex free electrons from its surroundings.

In the end it settles down into an ordinary unelectrified atom of an element whose atomic number is less by two than that of the original atom. Thus radium, which has the atomic number 88, sends out α-particles and becomes niton, with atomic number 86.

A substance may, however, be radio-active by sending out β-rays instead of α-rays. Some substances send out one kind, some the other, a very few can send out either, and can thus give rise to two different products of disintegration. The β-rays are simply very swiftly moving electrons, the most swiftly moving matter known to us; they attain velocities which reach very nearly the velocity of light. They have been known to travel at the rate of about 297,000 kilometres a second, while light covers 300,000 kilometres a second. (A kilometre is about five-eighths of a mile.) The velocity of light is a theoretical limit, which cannot be attained by anything material, so that we have in β-particles velocities about as great as we can ever expect to find in nature.

Since radio-activity always gives rise to a new element, and since the element is determined by its nucleus, the β-particles as well as the α-particles must come out of the nucleus.

RADIO-ACTIVITY 127

Since the β-particles are electrons, this shows that the nucleus of a radio-active element must contain electrons. This is to be expected in all elements except hydrogen, because the atomic weight increases about twice as fast as the atomic number, so that the atomic number (which is the net charge in the nucleus) must be the result of a number of hydrogen nuclei about twice as great as the atomic number and a number of electrons about equal to the atomic number. This is not always exactly true, but at any rate it is likely to be a first approximation. There is therefore no reason to be surprised by the fact that electrons come out of the nuclei of radio-active elements.

When an electron comes out of the nucleus of an atom, it increases the net charge in the nucleus by one, and therefore increases the atomic number by one. Thus it is possible for the atomic number to be increased by the *loss* of something from the nucleus, provided what is lost is an electron. But although the atomic number is increased, the atomic weight is not. The electron weighs so little that its loss makes no appreciable difference to the atomic weight ; moreover, since the net charge on the nucleus is increased by one, the atom will secure another planetary electron as soon as possible. Thus in the end the effect of a radio-active change by the emission of β-particles ought to be to leave the atomic weight unchanged while increasing the atomic number by one. The result of one emission of α-particles followed by two of β-particles is thus to deprive the nucleus of a helium nucleus and two electrons, without, in the end, changing the atomic number. Thus uranium has two very stable forms, called uranium I and uranium II. Uranium I is the great-grandfather of uranium II. Uranium I, by means of α-rays, gives rise to Uranium X_1, of which half decays in a little less than 24 days. Uranium X_1 gives rise to Uranium X_2 by means of β-rays; half of Uranium X_2 decays in just over a minute. Uranium X_2, by means of β-rays again, gives rise to Uranium II. Uranium I and II both have atomic number 92; Uranium X_1 has atomic number 90, and Uranium X_2 has atomic number 91. Thus the two stable forms of uranium have the same atomic number, although they have different atomic weights (238 and 234). Again, the various radio-active series all end in some form of lead; the three forms are called respectively radium-lead, actinium-lead, and thorium-lead, after their respective ancestors.

RADIO-ACTIVITY

These all have the same atomic number as ordinary lead (82), but their atomic weights differ. Ordinary lead has the atomic weight 207.2; radium-lead, 206.0; thorium-lead, 207.9. It is probable, however, that ordinary lead is a mixture of two or more kinds of lead, and perhaps this is also the case with what counts as thorium-lead. The reason for this view is that it is now probable that every perfectly pure element has an atomic weight which is almost exactly an integer.

When two elements have the same atomic number, they are called "isotopes." Apart from radio-activity, the only discoverable property in which isotopes differ is atomic weight. They have the same net charge in the nucleus, and therefore the same number of planetary electrons, and the same possible orbits of the electrons. Consequently their chemical properties are the same, their optical spectra are the same, and even their X-ray spectra are the same. All this is as it should be according to theory. It is no wonder that the existence of isotopes remained so long unknown. They first became known through observations of radio-active products. But it has lately become known, through the work of F. W. Aston, that there are many isotopes in regions of the periodic table where radio-activity can hardly be supposed to take place. Aston found methods, in a gas containing atoms of different weights, by which he separated the heavier and lighter atoms; he thus obtained two pure gases out of a mixture which had hitherto been wrongly supposed to be pure. The result was to show that atomic weights are very approximately integers in many cases in which this was thought not to be the case. Thus neon, which has the atomic weight 20.2, is found to consist of a mixture of two gases, one having atomic weight 20, the other 22. Chlorine, which has the atomic weight 35.46, is a mixture of two kinds having atomic weights 35 and 37 respectively. Krypton turns out to consist of as many as six isotopes; Xenon, of seven, two of which however are more or less doubtful.

It is a curious fact that in radio-activity the particles thrown off by the nucleus consist always either of electrons or of helium nuclei. Not only do we never find nuclei of heavier elements than helium thrown off, but we never find hydrogen nuclei. This is surprising, and as yet no adequate explanation has been found. What is to be said on this subject belongs to our next chapter.

The energy displayed in radio-activity is colossal. It shows that within the nucleus of the atom enormous forces are concentrated. This is not surprising when we consider that an atom as a whole is very minute, and that the nucleus of an atom is enormously smaller than the whole atom; that within the nucleus of uranium about 238 hydrogen nuclei and about 146 electrons are packed together; and that these attract or repel each other with a force which increases as the square of the distance diminishes. The energy involved is shown by the incredible swiftness of α-particles and β-particles. To make a body move with the velocity of light would require a strictly infinite amount of energy, and is therefore impossible, not only in practice but in theory. Therefore to make even so tiny a body as an electron move with a velocity 99 per cent. of that of light requires a very great amount of energy. Before the theory of relativity, the kinetic energy of a moving particle was taken to be half the mass multiplied by the square of the velocity; now-a-days, this has to be changed to allow for the increase of mass with velocity. In the case of a velocity 99 per cent. of that of light the mass is increased about seven-fold, and the kinetic energy in the same proportion. The energy of the α-particles, owing to their greater mass, is even more than that of the β-particles. As Sommerfeld says:

"The sources of energy which are thus disclosed to the external world are of quite a different order of magnitude from the energies of other physical and chemical processes. They bear witness what powerful forces are active in the interior of atoms (the nuclei). This world of the interior of the atom is in general closed to the outer world; it is not influenced by conditions of temperature and pressure which hold outside; it is ruled by the law of probability, of spontaneous chance which cannot be influenced. Only exceptionally a door opens, which leads from the inner world of the atom into the outer world; the α or β-rays which come out when this happens are envoys from a world which is otherwise closed to us."*

In the next chapter, we shall give an outline of the few facts which can be ascertained about this small fierce world in the nucleus of the atom. Optical spectra have told us about the outer electrons, X-rays about the inner rings of electrons; about the nucleus itself we know very little except what can be learnt from radio-activity.

*Op. cit. p. 109.

CHAPTER XI

THE STRUCTURE OF NUCLEI

The facts known about the structure of nuclei are not sufficient to enable us to be very definite as to the way in which they are built up. Let us begin by setting forth the facts, and then see what is to be inferred in the way of theory.

The most important facts for our purpose are those of radio-activity, which we considered in the previous chapter. We know from these facts that the nuclei of certain heavy atoms contain helium nuclei and electrons; that the loss of a helium nucleus diminishes the atomic number by two and the atomic weight by four, while the loss of an electron increases the atomic number by one and has no appreciable effect upon the atomic weight. We know also that it is possible for several different elements to exist with the same atomic number, but different atomic weights; in radio-activity these can be seen in process of formation, but they are found to exist among lighter atoms which show no discoverable trace of radio-activity. It is possible that radio-activity, in a very slight degree, exists among elements which appear to us perfectly stable ; the amount of it may be so small that we cannot hope to detect it. This may be the reason for the existence of isotopes ; but there is at present no positive evidence in favour of this view.

Another fact of immense importance has been experimentally established by Rutherford. Some elements, but not others, when submitted to a very intense bombardment by α-particles, give off rays which are found to be hydrogen. The element in which this result has been established with the greatest certainty is nitrogen (atomic number 17). But it is also fairly certain as regards a number of elements—broadly speaking, those which have odd atomic numbers. Rutherford is led to the conclusion that hydrogen nuclei can be detached from the nuclei of other elements, unless their atomic weight is a multiple of 4, which is the atomic weight of helium. This, together with the fact that in radio-activity helium nuclei, but not hydrogen nuclei, are thrown off, leads irresistibly to the view that

THE STRUCTURE OF NUCLEI

every nucleus is composed, as far as it can be, of helium nuclei. Thus, phosphorus, which has the atomic weight 31, may be supposed to consist of seven helium nuclei, each having atomic weight 4, and three hydrogen nuclei, each having atomic weight 1. The three hydrogen nuclei could, with luck, be detached by bombardment, but the helium nuclei are to be regarded as incapable of being destroyed by an α-particle, so that if they are detached they are detached as wholes. Consequently, when the atomic weight divides by 4, the nucleus can be supposed to consist wholly of helium nuclei, and there will be no odd hydrogen nuclei to be broken off. It is impossible to know whether Rutherford's bombardment breaks off helium nuclei, because they could not be distinguished from his projectiles, which are also helium nuclei.

The atomic weight, which we have hitherto found less important than the atomic number, is of course of the greatest significance when we are considering the structure of the nucleus. The weight of an electron is so small as to be negligible in comparison with that of the nucleus, even in the case of hydrogen, so that the weight of the atom is, to all intents and purposes, the weight of the nucleus. If we take the weight of the helium atom to be 4, the weight of the hydrogen atom is just over 1. The explanation of the fact that it is not exactly 1 is very interesting, and we shall return to it shortly. The weight of every other atom, in view of Aston's work on isotopes, is apparently a whole number, as nearly as our measurements can determine. Roughly speaking, the atomic weight is about double the atomic number. This is true exactly in the following cases (making, in some cases, inferences allowed by Aston's work) :

	Helium	Carbon	Oxygen	Neon	Sulphur	Calcium
Atomic number	2	6	8	10	16	20
Atomic weight	4	12	16	20	32	40

After this, the atomic weight is always more than double the atomic number. It will be seen that the above elements all have even atomic numbers and have atomic weights which divide by 4. We may therefore regard their nuclei as composed wholly of helium nuclei.

In the case of elements which have odd atomic numbers, there is only one instance, nitrogen, in which the atomic weight (14) is just double the atomic number (7). In this case, we may suppose that the nucleus consists of three helium nuclei and two hydrogen nuclei. In other cases, in the early part of

THE STRUCTURE OF NUCLEI 137

the periodic table, the atomic weight is greater by one than the double of the atomic number. Thus phosphorus has the atomic number 15, and the atomic weight 31. The same is true of the other early elements with odd atomic numbers, except nitrogen. (From the 21st element onward the atomic weight is larger than it would be by this rule.) The inference is that the nuclei of atoms which have odd atomic numbers usually consist of an adequate number of helium nuclei together with three hydrogen nuclei. The peculiarity of nitrogen is perhaps connected with the fact that Rutherford found it the easiest element from which to detach hydrogen nuclei.

The fact that the atomic weights are whole numbers, together with the facts of radioactivity and of Rutherford's bombardment, lead irresistibly to the conclusion that the weight of an atom is due to helium nuclei and hydrogen nuclei which exist together in its nucleus. The overcrowding in the nucleus of a heavy atom must be something fearful. Radium C, which emits the α-particles that Rutherford used in his experiments, has a nucleus whose radius is about three million-millionths of a centimetre (about one million-millionth of an inch). Its atomic number is

83 and its atomic weight is 214. This means that in this tiny space it must contain 53 helium nuclei and 2 hydrogen nuclei; it must also (as we shall see in a moment) contain 131 electrons. It is no wonder that helium nuclei and electrons move fast when radio-activity liberates them from this slum.

The β-rays show that the nucleus of an atom contains electrons. This appears also from the fact that the atomic number (which represents the net charge of the nucleus) is less than the atomic weight, which represents the gross positive charge. (Each hydrogen nucleus contributes one unit of positive charge.) The difference between the atomic weight and the atomic number represents the number of electrons there must be in the nucleus, in order to bring its net charge down to the atomic number. In this argument, however, we have assumed that the helium nucleus itself consists of four hydrogen nuclei and two electrons. We have still to examine the reasons in favour of this view.

There is no experimental evidence that a helium nucleus can be broken up into hydrogen nuclei and electrons. Radio-activity and Rutherford's bombardments show that the helium nucleus is very stable, and that no

THE STRUCTURE OF NUCLEI 139

known process will disintegrate it. Nevertheless it is believed by all students of the subject that the helium nucleus consists of four hydrogen nuclei and two electrons. There is first of all the argument from the atomic weight : the weight of the helium atom is so nearly four times the weight of the hydrogen atom that we cannot bring ourselves to attribute this fact to chance. But why is it not *exactly* four times the weight of a hydrogen atom ? If we take the weight of a helium atom as 4, that of a hydrogen atom is not 1, but 1.008. According to every-day notions, this would be impossible if a helium nucleus consisted of four hydrogen nuclei. (The electrons may be ignored, as their contribution to the weight is negligible.) We are used to thinking that if we place four pound weights in a scale, they will weigh four pounds. This, however, is only approximately true. In ordinary cases it is so nearly true that we could never discover the error experimentally; but in extraordinary cases, such as the helium nucleus, it may be sufficiently untrue for our measurements to be able to detect the difference.

It used to be thought that the mass of a body (which is the scientific conception that replaces the popular conception of weight) could be defined as the " quantity of matter." But Einstein has revolutionized the conception of mass, as well as all the other elementary conceptions of physics. Mass is now absorbed into energy, and the mass of a body is not by any means always constant.[*] A system of electrons and hydrogen nuclei may have different amounts of energy in different arrangements; when the system passes from an arrangement with more energy to one with less, the energy it loses is radiated into the surrounding medium, in the sort of way with which we became familiar when we were considering the spectrum of hydrogen. When the system loses energy it also loses mass. The loss of mass is very small compared to the loss of energy; it is obtained by dividing the loss of energy by half the square of the velocity of light, which is enormous. When the system has arranged itself in a shape in which its energy is diminished, it can only go back to its former shape if the lost energy is supplied from outside. Therefore the shapes involving least intrinsic energy are the most stable. This is what we must

[*] This subject of the variability of mass will be resumed in Chapter XIII.

THE STRUCTURE OF NUCLEI

suppose to happen when four hydrogen nuclei and two electrons come together to make a helium nucleus. They arrange themselves in a configuration in which their energy is less than when they were separated; the loss of energy can be inferred from the loss of mass (or weight, to speak popularly), and is got by multiplying this loss of mass by half the square of the velocity of light. This represents an enormous amount of energy. Sommerfeld calculated that it is about 10 million times greater than the amounts involved in chemical combinations (for instance, in combustion). The helium nucleus could only be disintegrated by supplying this amount of energy from outside, which does not happen in any known natural process. Thus the loss of weight in the helium atom is accounted for, and by the same argument the extreme stability of the helium nucleus is explained.

It is clear that, for the sake of unity and simplicity, it is desirable, if possible, to regard the helium nucleus as consisting of hydrogen nuclei and electrons. If we do not do so, we shall have to admit the helium nucleus as a third ultimate constituent of matter, having, by a strange coincidence, just twice the electric charge and four times the amount of matter that exists in the hydrogen nucleus. It must be admitted that this is a possible hypothesis; there are no known facts that prove it to be false. But until we are forced to adopt it, we shall prefer the simpler view that the helium nucleus is complex, like every other except hydrogen, and that its relations of mass and charge to the hydrogen atom are not a lucky fluke. Everything known about nuclei is consistent with the hypothesis that they are composed of hydrogen nuclei and electrons. The evidence that they consist of hydrogen nuclei, electrons, and helium nuclei is overwhelming; the further step, which dissolves the helium nucleus, is more or less hypothetical, but it is a step which we may take with a reasonable assurance that it will prove justified. The study of nuclei is still in its infancy, but is likely to make rapid advances in the near future. Meanwhile, we may assume, though not with complete certainty, that all matter consists of hydrogen nuclei and electrons, which are therefore the only " elements " in the strict sense of the word. Whether these two will ultimately prove to be modifications of some one more fundamental substance, it is quite impossible to say. For the present, they

THE STRUCTURE OF NUCLEI

represent the frontier of scientific knowledge, and what lies beyond is as yet mere speculation.

As to the way in which the four hydrogen nuclei and the two electrons are arranged in the helium atom, mathematical considerations ought to be able to give us information, but so far they have not given much. One model which is suggestive is the following: Imagine a somewhat primitive wheel, with four spokes, and an axle that sticks out some distance to either side. Place the two electrons at the ends of the axle, and the four hydrogen nuclei at the ends of the spokes, and imagine the wheel going round with suitable velocity. (The wheel and spokes and axle are of course imaginary, and are only intended to illustrate the relative positions of the nuclei and electrons.) This gives a configuration which has a certain degree of stability, and a flattish shape which is indicated by a certain amount of experimental evidence. It seems however, that the degree of stability in this model is less than that required to account for the fact that no known process will disintegrate a helium nucleus. There is also a difficulty as regards the size of the helium nucleus. Taking our model and applying the quantum theory to the revolutions of the hydrogen nuclei, we can determine the radius of the circle in which they move as we determined the minimum orbit in the hydrogen atom. The result is that the size of the radius should be about 5 million-millionths of a centimetre. This is about seventeen times too large, according to Rutherford's experimental evidence. It is possible, nevertheless, that our model may be right, because the forces between electrons and hydrogen nuclei may obey different laws, at such very tiny distances, from those which they obey at ordinary distances. We may hope to know more on this subject at no distant date, but for the present we must remain in doubt.

CHAPTER XII

THE NEW PHYSICS AND THE WAVE-THEORY OF LIGHT.

In the physics of the atom, as it has become in modern times, everything is atomic, and there are sudden jumps from one condition to another. The electron and the hydrogen nucleus are the true " atoms " both of electricity and of matter. According to the quantum theory, there are also atomic quantities, not of energy as was thought when the theory was first suggested, but of what is called " action." The word " action," in physics, has a precise technical meaning; it may be regarded as the result of energy operating for a certain time. Thus if a given amount of energy operates for two seconds there is twice as much action as if it operated for one second; if it operates for a minute there is 60 times as much action, and so on. If twice the amount of energy operates for a second, there is again twice as much action, and so on. If the energy which is operating is variable, and we wish to estimate its action during a given time, we divide the time into a number of little bits, during each of which the energy will vary so little that it may be regarded as constant ; we then multiply the energy during each little interval of time by the length of the interval, and add up for all the intervals. As we make the intervals smaller and more numerous, the result of our addition approaches nearer and nearer to a certain limit ; this limit we define as the total action during the total period of time concerned. Action is a very important conception in physics ; from the point of view of theory it is more important than energy, which has been deposed from its eminence by the theory of relativity. Planck's quantum h is of the nature of action ; thus the quantum theory amounts to saying that there are atoms of action.

So long as we confine ourselves to what goes on in matter, this theory is self-consistent and explains the facts, nor is it easy to suppose that any theory which was not atomic would explain the facts. But when we come to what goes on in " empty space," or in the " aether " we find ourselves in difficulties if we adhere to the quantum theory. Consider what happens

THE NEW PHYSICS

when a wave of light is sent out by an atom, with only one quantum of action in each period. The wave spreads out in all directions, growing fainter as it goes on, like a ripple when a stone is dropped into a pond. The evidence that light consists of waves remains quite unshaken; it is derived from the phenomena of interference and diffraction. As to interference, a few words may be necessary. If two sets of waves are travelling more or less in the same direction, if their crests come together they will grow bigger but if the crest of one comes in the same place as the trough of the other, they will neutralize each other. Now it is possible to arrange two sets of light-waves so that in some places their crests come together, while in others the crest of one covers the trough of the other. When this is done, we get a lattice pattern of alternate light and darkness, light where the waves reinforce each other, and darkness where they neutralize each other. If light consisted of particles travelling, and not of waves, this phenomenon, which is called "interference," could not take place.

The difficulties which arise for the quantum-theory out of the phenomenon of interference have been forcibly stated by Jeans in the following paragraphs :*

"If light occurred only in quanta, interference could only occur at a point at which two or more quanta existed simultaneously. If the light were sufficiently feeble the simultaneous occurrence of two quanta at any point ought to be a very rare occurrence, so that all phenomena, such as diffraction patterns, which depend on interference, ought to disappear as the quantity of light present is reduced. Taylor has shown that this is not the case; he reduced the intensity of his light to such an extent that an exposure of 2,000 hours was necessary to obtain a photograph, and yet obtained photographs of diffraction patterns in which the alternation of light and dark appeared with undiminished sharpness. In Taylor's experiment the intensity of light was . . . about one light-quantum per 10,000 cubic centimetres, so that if the quanta had been concentrated nothing of the nature of a diffraction could possibly have been observed.

"Thus it appears that there is no hope of reconciling the undulatory theory of light with the quantum-theory by regarding the

*Report on Radiation and the Quantum Theory, p. 87.

THE NEW PHYSICS 149

undulatory theory as being, so to speak, only statistically true when a great number of quanta are present. One theory cannot be the limit of the other in the sense in which the Newtonian mechanics is the limit of the quantum-mechanics, and we are faced with the problem of combining two apparently quite irreconciliable theories."

Other similar difficulties might be mentioned, but the difficulty of interference may suffice, since it is typical. It may be questioned whether the difficulty still exists when the quantum theory is stated in the form which it takes in Sommerfeld's work. We no longer have little parcels of energy; what we have is a property of periodic processes. It would not be accurate to state this property in the form : the total action throughout a complete period of any periodic process is h or an exact multiple of h. But although this statement would not be accurate, it gives, as nearly as is possible in non-mathematical language, a general idea of the sort of thing that is affirmed by the modern form of the quantum theory. In order to reconcile this principle with the facts about the diffusion of light, it is only necessary to avoid dividing the aether into imaginary particles. As the light-wave travels outward, so long as it meets no obstruction its energy remains constant, though it is more diffused, so that there is less of it in any given area of the wave-front. But while we remain in empty space, the wave must be treated as a whole, and the quantum theory must not be applied to separate little bits of it. The quantum-theory has to do, not with what is happening in a point at an instant, but with what happens to a periodic process throughout its whole period. Just as the period occupies a certain finite time, so the process occupies a certain finite space; and in the case of a light-wave travelling outward from a source of light, the finite space occupied by the process grows larger as it travels away from the source. For the purposes of stating the quantum principle, one period of a periodic process has to be treated as an indivisible whole. This was not evident at the time when Jean's report was written (1914), but has been made evident by subsequent developments. While it makes the quantum principle more puzzling, it also prevents it from being inconsistent with the known facts about light.

It must be confessed that the quantum principle in its modern form is far more astonishing and bewildering than is its older form.

It might have seemed odd that energy should exist in little indivisible parcels, but at any rate it was an idea that could be grasped. But in the modern form of the principle, nothing is said, in the first instance, about what is going on at a given moment, or about atoms of energy existing at all times, but only about the total result of a process that takes time. Every periodic process arranges itself so as to have achieved a certain amount by the time one period is completed. This seems to show that nature has a kind of foresight, and also knows the integral calculus, without which it is impossible to know how fast to go at each instant so as to achieve a certain result in the end. All this sounds incredible. No doubt the fact is that the principle has assumed a complicated form because it has forced its way through, owing to experimental evidence, in a science built upon totally different notions. The revolution in physical notions introduced by Einstein has as yet by no means produced its full effect. When it has, it is probable that the quantum principle will take on some simple and easily intelligible form. But it will only be easily intelligible to those who have gone through the labour of learning to think in terms of modern physical notions rather than in terms of the notions derived from common sense and embodied in traditional physics. In the last chapter of this book we shall try to indicate the sort of way in which this may affect the quantum principle.

It is necessary, however, to utter a word of warning, in case readers should accept as a dogmatic ultimate truth the atomic structure of the world which we have been describing, and which seems at present probable. It should not be forgotten that there is another order of ideas, temporarily out of fashion, which may at any moment come back into favour if it is found to afford the best explanation of the phenomena. The charge on an electron, the equal and opposite charge on a hydrogen atom, the mass of an electron, the mass of a hydrogen nucleus, and Planck's quantum, all appear in modern physics as absolute constants, which are just brute facts for which no reason can be imagined. The aether, which used to play a great part in physics, has sunk into the background, and has become as shadowy as Mrs. Harris. It may be found, however, as a result of further research, that the aether is after all what is really fundamental, and that electrons and

THE NEW PHYSICS

hydrogen nuclei are merely states of strain in the aether, or something of the sort. If so, the two "elements" with which modern physics operates may be reduced to one, and the atomic character of matter may turn out to be not the ultimate truth. This suggestion is purely speculative; there is nothing in the existing state of physics to justify it. But the past history of science shows that it should be borne in mind as a possibility to be tested hereafter. If the possibility should be realized, it would not mean that the present theory is false; it would merely mean that a new interpretation had been found for its results. Our imagination is so incurably concrete and pictorial that we have to express scientific laws, as soon as we depart from the language of mathematics, in language which asserts much more than we *mean* to assert. We speak of an electron as if it were a little hard lump of matter, but no physicist really means to assert that it is. We speak of it as if it had a certain size, but that also is more than we really mean. It may be something more analogous to a noise, which is spread throughout a certain region, but with diminishing intensity as we travel away from the source of the noise. So it is possible that an electron is a certain kind of disturbance in the aether, most intense at one spot, and diminishing very rapidly in intensity as we move away from this spot. If a disturbance of this sort could be discovered which would move and change as the electron does, and have the same amount of energy as the electron has and have periodic changes of the same frequency as those of the electron, physics could regard it as what an electron really is without contradicting anything that present-day physics means to assert. And of course it is equally possible that a hydrogen nucleus may come to be explained in a similar way. All this is, however, merely a speculative possibility; there is not as yet any evidence making it either probable or improbable. The only thing that is probable is that there will be such evidence, one way or other, before many years have passed.

CHAPTER XIII

THE NEW PHYSICS AND RELATIVITY

The theory of quanta and the theory of relativity have been derived from very different classes of phenomena. The theory of quanta is concerned with the smallest quantities known to science, the theory of relativity with the largest. Distances too small for the microscope are concerned in the theory of quanta; distances too large for the telescope are concerned in the theory of relativity. Relativity came, in the first instance, from astronomy and the study of the propagation of light in astronomical spaces, and its most noteworthy triumphs have been in regard to astronomical phenomena—the motion of the perihelion of Mercury, and the bending of light from the stars when it passes near the sun. The material of the quantum theory, on the contrary, is mainly derived from small quantities of very rarefied gases in laboratories, and from tiny particles running about in a vacuum as nearly perfect as we can make it. In the theory of relativity, 300,000 kilometres counts as a small distance; in the theory of quanta, a thousandth of a centimetre counts as infinitely great. The result of this divergence is that the two theories have been pursued by different investigators, because they required different apparatus and different methods. In this final chapter, we shall consider what bearing the two theories have on each other, and, in particular, whether there is anything in relativity that makes the theory of quanta seem less odd and irrational.

The theory of relativity, as every one knows, was discovered by Einstein in two stages, of which the first is called the special theory and the second the general theory. The first dates from 1905, the second from 1915. The first is not superseded by the second, but absorbed into it as a part. We shall not attempt to explain the theory of relativity, which has been done popularly (so far as is possible) in a multitude of books, and scientifically in two books which should be read by all who have sufficient mathematical equipment: Hermann Weyl's "Space, Time, Matter." and Eddington's "Mathematical

NEW PHYSICS AND RELATIVITY 157

Theory of Relativity." We are only concerned with the points where this theory touches the problem of atomic structure.

The special theory of relativity, as we have already seen, is relevant to the problems we have been considering at several points. It is relevant through its doctrine that mass, as measured by our instruments, varies with velocity, and is, in fact, merely a part of the energy of a body. It is part of the theory of relativity to show that the results of measurement, in a great many cases, do not yield physical facts about the quantities intended to be measured, but are dependent upon the relative motion of the observer and what is observed. Since motion is a purely relative thing, we cannot say that the observer is standing still while the object observed is moving; we can only say that the two are moving relatively to each other. It follows that any quantity which depends upon the motion of a body relatively to the observer cannot be regarded as an intrinsic property of the body. Mass, as commonly measured, is such a property; if the body is moving with a velocity which approaches that of light, its measured mass increases, and as the velocity gets nearer to that of light, the measured mass increases without limit. But this increase of mass is only apparent; it would not exist for an observer moving with the body whose mass is being measured. The mass as measured by an observer moving with the body is what counts as the true mass, and it is easily inferred from the measured mass when we know how the body concerned is moving relatively to ourselves. When we say that any two electrons have the same mass, or that any two hydrogen nuclei have the same mass, we are speaking of the true mass. The apparent mass of an electron which is shot out in the form of a β-ray may be several times as great as the true mass.

There are two other points where the variability of apparent mass is relevant in the theory of atoms. One concerns the "fine structure" and the analogy between the electron in a hydrogen atom and the planet Mercury; this was considered in Chapter VII. The other is the explanation of the fact that the helium nucleus is less than four times as heavy as the hydrogen nucleus, which concerned us in Chapter XI. On both these points, as we have seen, the theory of relativity provides admirably satisfactory explanations of facts which would otherwise remain obscure. Both, however, raise the question of the relativity of energy, which might be thought awkward for the quantum theory, because this theory uses the conservation of energy, and something merely relative to the observer cannot be expected to be conserved.

In elementary dynamics, as every one knows, energy consists of two parts, kinetic and potential. Ignoring the latter, let us consider the former. The kinetic energy depends upon the mass and the velocity, but the velocity depends upon the observer, and is not an intrinsic property of a body. The result is that energy has to be re-defined in the theory of relativity. It turns out that we can identify the energy of a body with its mass as measured by the observer (or, in ordinary units, with this mass multiplied by the square of the velocity of light). Although, for a particular body, this mass varies with the observer, its sum throughout the universe will be constant for a given observer, however he may be moving.*

In the theory of relativity, there are two kinds of variation of mass to be distinguished, of which so far we have only considered one

*Eddington, *op cit.*, pp. 30-32.

We have considered the change of measured mass (as we have called it) which is brought about by a change in the relative motion of the observer and the body whose mass is being measured. This is not a change in the body itself, but merely in its relation to the observer. It is this change which has to be allowed for in deducing from experimental data that all electrons have the same mass. We allow for it by means of a formula, which enables us to infer what we may call the " proper mass " of the body. This is the mass which it will be found to have by an observer who shares its motion. In all ordinary cases, in which we determine mass (or weight) by means of a balance, we and the body which we are weighing share the same motion namely that of the earth in its rotation and revolution ; thus weighing with a balance gives the " proper mass." But in the case of swiftly-moving electrons and α-particles we have to adopt other ways of measuring their mass, because we cannot make ourselves move as fast as they do ; thus in these cases we only arrive at the " proper mass " by a calculation. The " proper mass " is a genuine property of a body, not relative to the observer. As a rule, the proper mass is constant, or very nearly so, but it is not always strictly constant. When a body absorbs radiant energy, its proper mass is increased; when it radiates out energy, its proper mass is diminished. When four hydrogen nuclei and two electrons combine to form a helium nucleus, they radiate out energy. The loss of mass involved is loss of proper mass, and is quite a different kind of phenomenon from the variation of measured mass when an electron changes its velocity.

There is another point, not easy to explain clearly, and as yet amounting to no more than a suggestion, but capable of proving very important in the future. We saw that Planck's quantum h is not a certain amount of energy, but a certain amount of what is called "action." Now the theory of relativity would lead us to expect that action would be more important than energy. The reason for this is derived from the fact that relativity diminishes the gulf between space and time which exists in popular thought and in traditional physics. How this affects our question we must now try to understand.

Consider two events, one of which happens at noon on one day in London, while the other happens at noon the next day in Edinburgh. Common sense would say that there are two kinds of intervals between these two events, an interval of 24 hours in time, and an interval of 400 miles in space. The theory of relativity says that this is a mistake, and that there is only one kind of interval between them, which may be analysed into a space-part and a time-part in a number of different ways. One way will be adopted by a person who is not moving relatively to the events concerned, while other ways will be adopted by persons moving in various ways. If a comet were passing near the earth when our two events happened, and were moving very fast relatively to the earth, an observer on the comet would divide the interval of our two events differently between space and time, although, if he knew the theory of relativity, he would arrive at the same estimate of the total interval as would be made by our relativity physicists. Thus the division of the interval into a space portion and a time portion does not belong to the physical relation of the two events, but is something subjective, contributed by the observer. It cannot, therefore, enter into the correct statement of any law of the physical world.

The importance of this principle (which is supported by a multitude of empirical facts) is impossible to exaggerate. It means, in the first place, that the ultimate facts in physics must be events, rather than bodies in motion. A body is supposed to persist through a certain length of time, and its motion is only definite when we have fixed upon one way of dividing intervals between space and time. Therefore any physical statement in terms of the motions of bodies is in part conventional and subjective, and must contain an element not belonging to the physical occurrence. We have therefore to deal with events, whose relative positions, in the conventional space-time system that we have adopted, are fixed by four quantities, three giving their relations in space (e.g. east-and-west, north-and-south, up-and-down), while the fourth gives their relation in time. The true interval between them can be calculated from these, and is the same whatever conventional system we adopt ; just as the time-interval between two historical events would be the same whether we dated both by the Christian era or by the Mohammedan, only that the calculation is not so simple.

It follows from these considerations that, when we wish to consider what is happening in some very small region (as we have to do whenever we apply the differential or integral calculus), we must not take merely a small region of space, but a small region of space-time, i.e. in conventional language, what is happening in a small volume of space during a very short time. This leads us to consider, not merely the energy at an instant, but the effect of energy operating for a very short time ; and this, as we saw, is of the nature of action (in the technical sense). A quotation from Eddington * will help to make the point clear :

" After mass and energy there is one physical quantity which plays a very fundamental part in modern physics, known as *Action*. *Action* here is a very technical term, and is not to be confused with Newton's ' Action and Reaction.' In the relativity theory in particular this seems in many respects to be the most fundamental thing of all. The reason is not difficult to see. If we wish to speak of the continuous matter present *at* any particular point of space and time, we must use the term *density*. Density multiplied by volume in space gives us *mass* or, what appears to be the same thing, *energy*. But from our space-time point of view, a far more

*Space, Time and Gravitation, p. 147.

NEW PHYSICS AND RELATIVITY 165

important thing is density multiplied by a four-dimensional volume of space and time; this is *action*. The multiplication by three dimensions gives mass or energy; and the fourth multiplication gives mass or energy multiplied by time. Action is thus mass multiplied by time, or energy multiplied by time, and is more fundamental than either."

It is a fact which must be significant that action thus turns out to be fundamental both in relativity theory and in the theory of quanta. But as yet it is impossible to say what is the interpretation to be put upon this fact; we shall probably have to wait for some new and more fundamental way of stating the quantum theory.

There is one other respect in which some of the later developments of relativity suggest the possibility of answers to questions which have hitherto seemed quite unanswerable. Our theory, so far, leads us to brute facts which have to be merely accepted. We do not know why there are two kinds of electricity, or why opposite kinds attract each other while similar kinds repel each other. This dualism is one of the things which is intellectually unsatisfying about the present condition of physics. Another thing is the conflict between the discontinuous process by which energy is radiated from the atom into the surrounding medium, and the continuous process by which it is transmitted through the surrounding medium. Relativity throws very little light on these points, but there is another point upon which it throws at least a glimmer. We find it hard to rest content with the existence of unrelated absolute constants, such as Planck's quantum and the size of an electron, which, so far as we can see, might just as easily have had any different magnitude. To the scientific mind, such facts are a challenge, leading to a search for some way of inter-relating them and making them seem less accidental. As regards the quantum, no plausible suggestion has yet been made. But as regards the size of an electron, Eddington makes some suggestive observations, which, however, require some preliminary explanations.

We saw that, according to the theory of relativity, the interval between two events may be separated into a time-part and a space-part in various ways, all of which are equally legitimate, and each of which will seem natural to an observer who is moving suitably.

NEW PHYSICS AND RELATIVITY 167

The first effect of this is to diminish the sharpness of the distinction between space and time. But the distinction comes back in a new form. It is found that the interval between two events can, in some cases, be regarded as merely a space-interval; this will happen if an observer who is moving suitably would regard them as simultaneous. Whenever this does not happen, the interval can be regarded as merely a time-interval; this will be the case when an observer could travel so as to be present at both events. It takes eight minutes for light to travel from the sun to the earth, and nothing can travel faster than light; therefore if we consider some event which happens on the earth at 12 noon, any event which happens on the sun between 11.52 a.m. and 12.8 p.m. could not have happened in the presence of anything which was present at the event on earth at 12 noon. Events happening on the sun during these 16 minutes have an interval from the event on earth which will, for a suitable observer seem to be a spatial separation between simultaneous events; such intervals are called space-like. Events happening earlier or later than these 16 minutes will be separated from the event on earth at noon by an interval which would appear to be purely temporal to an observer who had spent the interval in travelling from the sun to the earth, or vice versa as the case may be; such intervals are called time-like. Two parts of one light-ray are on the borderland between time-like and space-like intervals, and in fact the interval between them is zero. But in all other cases there is a separation which is either time-like or space-like, and in this way we find that there is still a distinction between what is to be called temporal and what is to be called spatial, though the distinction is different from that of every-day life.

For reasons which we cannot go into, Einstein and others have suggested that the universe has a " curvature," so that we could theoretically go all round it and come back to our starting-point, in the sort of way in which we go round the earth. All the way round the universe, in that case, must be a certain length, fixed in nature. Eddington suggests that some relation will probably be found between this, the greatest length in nature, and the radius of the electron, which is the least length in nature. As he humorously puts it: " An electron could never decide how large it ought to be unless there existed some length independent of itself for it to compare itself with."

He goes on to make another application of this principle, which is suggestive, though perhaps not intended to be treated too solemnly. The curvature of the universe, if it exists, is only in space, not in time. This leads him to say :*

"By consideration of extension in time-like directions we obtain a confirmation of these views which is, I think, not entirely fantastic. We have said that an electron would not know how large it ought to be unless there existed independent lengths in space for it to measure itself against. Similarly it would not know how long it ought to exist unless there existed a length in time for it to measure itself against. But there is no radius of curvature in a time-like direction; so the electron does *not* know how long it ought to exist. Therefore it just goes on existing indefinitely."

But even if the size of an electron should ultimately prove, in this way, to be related to the size of the universe, that would leave a number of unexplained brute facts, notably the quantum itself, which has so far defied all

*The Mathematical Theory of Relativity, p. 155.

attempts to make it seem anything but accidental. It is *possible* that the desire for rational explanation may be carried too far. This is suggested by some remarks, also by Eddington, in his book on " Space, Time and Gravitation " (p. 200). The theory of relativity has shown that most of traditional dynamics, which was supposed to contain scientific laws, really consisted of conventions as to measurement, and was strictly analogous to the " great law " that there are always three feet to a yard. In particular, this applies to the conservation of energy. This makes it plausible to suppose that every apparent law of nature which strikes us as reasonable is not really a law of nature, but a concealed convention, plastered on to nature by our love of what we, in our arrogance, choose to consider rational. Eddington hints that a real law of nature is likely to stand out by the fact that it appears to us irrational, since in that case it is less likely that we have invented it to satisfy our intellectual taste. And from this point of view he inclines to the belief that the quantum-principle is the first real law of nature that has been discovered in physics.

This raises a somewhat important question : Is the world " rational," i.e., such as to conform to our intellectual habits? Or is it "irrational," i.e., not such as we should have made it if we had been in the position of the Creator? I do not propose to suggest an answer to this question.

APPENDIX

BOHR'S THEORY OF THE HYDROGEN SPECTRUM

The mathematics involved in this theory is so simple that only a very slight acquaintance with elementary dynamics is required in order to understand it.

Let us consider an electron revolving in a circle about the nucleus. Let m be the mass of the electron, a the radius of its orbit, ω its angular velocity. Also let e be the (negative) charge on the electron and the (positive) charge on the nucleus.

Then according to elementary dynamics, the centrifugal force of the electron in its orbit is

$$m\,a\,\omega^2$$

while the force attracting it to the nucleus is

$$\frac{e^2}{a^2}$$

by Coulomb's Law. These two must be equal, so that

$$m\,a\,\omega^2 \;=\; \frac{e^2}{a^2} \tag{1}$$

BOHR'S THEORY

So far, we have been proceeding on traditional lines. But we come now to the application of the quantum theory.

The kinetic energy of the electron is $\frac{1}{2} m a \omega^2$; the potential energy is $-\frac{e^2}{a}$. In virtue of the above equation, $\frac{e^2}{a}$ is double $\frac{1}{2} m a^2 \omega^2$, so that the total energy is equal to the kinetic energy with its sign changed. The impulse corresponding to ω is $m a^2 \omega$, and this has to be taken round one complete circuit of the orbit. This yields the value $2\pi m a^2 \omega$, which must be put equal to a multiple of h, say $n h$, where n is an integer. Thus we have the equation

$$2 \pi m a^2 \omega = n h \qquad (2)$$

Now m and e and h are known; thus (1) and (2) determine a and ω as soon as n is fixed. We have

$$a = \frac{n^2 h^2}{4 \pi^2 m e^2}, \quad \omega = \frac{8 \pi^3 m e^4}{n^3 h^3}$$

The smallest possible orbit is got by putting $n = 1$; thus its radius is a_1, where

$$a_1 = \frac{h^2}{4 \pi^2 m^2 e}$$

The next possible radius is

$$a_2 = \frac{4 h^2}{4 \pi^2 m e^2} = 4 a_1$$

The kinetic energy in the nth orbit is

$$\tfrac{1}{2} m a^2 \omega^2 = \tfrac{1}{2} m \left(\frac{2\pi e^2}{nh}\right)^2$$

Since the total energy is the kinetic energy with its sign changed, the loss of energy in passing from the kth to the nth orbit is

$$\tfrac{1}{2} m \left(\frac{2\pi e^2}{h}\right)^2 \left(\frac{1}{n^2} - \frac{1}{k^2}\right)$$

If this transition is to give rise to a wave of frequency ν, we must have

$$\tfrac{1}{2} m \left(\frac{2\pi e^2}{h}\right)^2 \left(\frac{1}{n^2} - \frac{1}{k^2}\right) = h\nu$$

by the principle of quanta. That is to say, ν is given by the equation

$$\nu = m \, \frac{2\pi^2 e^4}{h^3} \left(\frac{1}{n^2} - \frac{1}{k^2}\right)$$

If c is the velocity of light, this gives a wave-number ν/c. Now the empirical formula for the wave-numbers of the lines of the hydrogen spectrum is

$$R \left(\frac{1}{n^2} - \frac{1}{k^2}\right)$$

where R is Rydberg's constant. This shows that, if our theory is right, we ought to have

$$R = \frac{2\pi^2 m e^4}{h^3 c}$$

By substituting the observed values for m, e, h and c, it is found that this equation is satisfied. This was perhaps the most sensational evidence in favour of Bohr's theory when it was first published.

TABLE OF THE ELEMENTS

Atomic Number	Element	Atomic Weight
First Period		
1	Hydrogen	1.008
2	Helium	4.00
Second Period		
3	Lithium	6.94
4	Beryllium	9.01
5	Boron	10.82
6	Carbon	12.00
7	Nitrogen	14.01
8	Oxygen	16.00
9	Fluorine	19.0
10	Neon	20.2
Third Period		
11	Sodium	23.00
12	Magnesium	24.32
13	Aluminium	27.1
14	Silicon	28.3
15	Phosphorus	31.04
16	Sulphur	32.06
17	Chlorine	35.46
18	Argon	39.88
Fourth Period		
19	Potassium	39.10
20	Calcium	40.07
21	Scandium	45.1
22	Titanium	48.1
23	Vanadium	51.0
24	Chromium	52.0
25	Manganese	54.93
26	Iron	55.84
27	Cobalt	58.97
28	Nickel	58.68
29	Copper	63.57
30	Zinc	65.37
31	Gallium	69.9
32	Germanium	72.5
33	Arsenic	74.96
34	Selenium	79.2
35	Bromine	79.92
36	Krypton	82.92
Fifth Period		
37	Rubidium	85.45
38	Strontium	87.63
39	Yttrium	88.7
40	Zirconium	90.6
41	Niobium	93.5
42	Molybdenum	96.0
43		
44	Ruthenium	101.7
45	Rhodium	102.9
46	Palladium	106.7
47	Silver	107.88
48	Cadmium	112.40
49	Indium	114.8
50	Tin	118.7
51	Antimony	120.2
52	Tellurium	127.5
53	Iodine	126.92
54	Xenon	130.2

Atomic Number	Element	Atomic Weight		Atomic Number	Element	Atomic Weight
Sixth Period				74	Tungsten	184.0
				75		
55	Caesium	132.81		76	Osmium	190.9
56	Barium	137.37		77	Iridium	193.1
57	Lanthanum	139.0		78	Platinum	195.2
58	Caerium	140.25		79	Gold	197.2
59	Praseodymium	140.9		80	Mercury	200.6
60	Neodymium	144.3		81	Thallium	204.4
61				82	Lead	207.20
62	Samarium	150.4		83	Bismuth	209.00
63	Europium	152.0		84	Polonium	210.0
64	Gadolinium	157.3		85		
65	Terbium	159.2		86	Niton	222.0
66	Dysprosium	162.5		*Seventh Period*		
67	Holmium	163.5				
68	Erbium	167.7		87		
69	Thulium	168.5		88	Radium	226.0
70	Ytterbium	173.5		89	Actinium	226
71	Lutecium	175.0		90	Thorium	232.12
72	Hafnium	178		91	Protoactinium	230
73	Tantalum	181.5		92	Uranium	238.2

Rare Earths: 61–71